Praise for *G*

"Owen . . . makes a convincing case that Manhattan, Hong Kong, and large, old European cities are inherently greener than less densely populated places because a higher percentage of their inhabitants walk, bike, and use mass transit than drive; they share infrastructure and civic services more efficiently; they live in smaller spaces and use less energy to heat their homes (because those homes tend to share walls); and they're less likely to accumulate a lot of large, energy-sucking appliances. Pugnacious and contrarian, this book has a lot of fun at the expense of sentimental pastoralists, high-minded environmentalists, and rich people trying to buy their way into higher green consciousness."

—*The New York Times Book Review*

"The deservedly respected journalist David Owen spent a lot of time in recent years patrolling the environmental beat, doing research for the excellent book we now have before us. Owen's style . . . is cool, understated, and witty; it does not appear to be in his nature to be alarmist. But this is a thoroughly alarming book, perhaps all the more so because Owen is so matter-of-fact: The facts alone are so discouraging that no rhetorical flourishes are necessary to underscore their urgency. No one can wave a wand and turn the environmental disasters of some cities across the country into instant Manhattans, with tall apartment buildings densely situated, efficient mass transit and zillions of pedestrians. The much more likely prospect is that we will just keep stumbling along, indulging ourselves and closing our eyes to reality until it crashes in on us—sooner rather than later—with highly unpleasant and probably calamitous consequences."

—*The Washington Post*

continued . . .

"David Owen advances the provocative argument that the asphalt jungle is greener than the places where most Americans live. A hard-hitting book that punctures many eco-balloons."

—Witold Rybczynski, author of *City Life* and Meyerson Professor of Urbanism, University of Pennsylvania

"The future of our planet may be uncertain, but one thing is clear: David Owen is going to generate significant heat with *Green Metropolis,* his provocative manifesto that inverts many of our sacred assumptions about environmentalism. . . . His book mounts a passionate, fact-studded case for 'the environmental advantages of Manhattan-style urban density.' Fascinating and thought-provoking . . . *Green Metropolis* offers high-test fuel for discussion." —NPR

"Seminal . . . to be read by anyone concerned with the true meaning of sustainability. In chapter after chapter, Owen punctures our myths surrounding the green movement with laser-guided precision in the hopes of clearing the air. His method is provocative . . . with irreverent lucidity, he forces us to abandon unfounded beliefs, allowing the sustainability movement to evolve and mature, one realization and one city at a time." —*Architectural Record*

"David Owen always delights with his elegant insights and his challenges to conventional thinking. In this book, he does so again by puncturing the myth of ecological Arcadia and reminding us why living in cities is the best way to be green. It's a triumph of clear thinking and writing."

—Walter Isaacson, author of *Steve Jobs*

"A bracing, important work of contrarian truth-telling."

—Kurt Andersen, author of *The Real Thing* and *Heyday*

"While the conventional wisdom condemns it as an environmental nightmare, Manhattan is by far the greenest place in America, argues this stimulating eco-urbanist manifesto. Owen's lucid, biting prose crackles with striking facts that yield paradigm-shifting insights. The result is a compelling analysis of the world's environmental predicament that upends orthodox opinion and points the way to practical solutions." —*Publishers Weekly*

"*New Yorker* writer Owen lays out a simple plan to address our environmental crisis—live smaller, live closer, and drive less. He presents a convincing argument. . . . He effectively connects the dots among oil, cars, public transportation, ethanol, rising food prices, and the role of plastic in modern life. VERDICT: Owen's engaging, accessible book challenges the idea of green and urban living."

—*Library Journal*

"Want to reduce your carbon footprint and save the planet? Move to Manhattan. Owen assembles useful facts, some of them sure to be surprises even for the most learned of NYC boosters. . . . [He] provides a dogged, contrarian argument that scores some good points." —*Kirkus Reviews*

ALSO BY DAVID OWEN

Green Metropolis 2009

Sheetrock & Shellac 2006

Copies in Seconds 2004

The First National Bank of Dad 2003

Hit & Hope 2003

The Chosen One 2001

The Making of the Masters 1999

Around the House
(also published as Life Under a
Leaky Roof) 1998/2000

Lure of the Links (coeditor) 1997

My Usual Game 1995

The Walls Around Us 1991

The Man Who Invented Saturday
Morning 1988

None of the Above 1985

High School 1981

The Conundrum

How Scientific Innovation, Increased Efficiency, and Good Intentions Can Make Our Energy and Climate Problems Worse

DAVID OWEN

RIVERHEAD BOOKS
New York

RIVERHEAD BOOKS
Published by the Penguin Group
Penguin Group (USA) Inc.
375 Hudson Street, New York, New York 10014, USA
Penguin Group (Canada), 90 Eglinton Avenue East, Suite 700, Toronto, Ontario M4P 2Y3, Canada (a division of Pearson Penguin Canada Inc.)
Penguin Books Ltd., 80 Strand, London WC2R 0RL, England
Penguin Group Ireland, 25 St. Stephen's Green, Dublin 2, Ireland (a division of Penguin Books Ltd.)
Penguin Group (Australia), 250 Camberwell Road, Camberwell, Victoria 3124, Australia (a division of Pearson Australia Group Pty. Ltd.)
Penguin Books India Pvt. Ltd., 11 Community Centre, Panchsheel Park, New Delhi—110 017, India
Penguin Group (NZ), 67 Apollo Drive, Rosedale, Auckland 0632, New Zealand (a division of Pearson New Zealand Ltd.)
Penguin Books (South Africa) (Pty.) Ltd., 24 Sturdee Avenue, Rosebank, Johannesburg 2196, South Africa

Penguin Books Ltd., Registered Offices: 80 Strand, London WC2R 0RL, England

Book design by Tiffany Estreicher

First Riverhead trade paperback edition: February 2012

Library of Congress Cataloging-in-Publication Data

Owen, David, date.
The conundrum : how scientific innovation, increased efficiency, and good intentions can make our energy and climate problems worse / David Owen. — 1st Riverhead trade pbk. ed.
p. cm.
Includes bibliographical references.
ISBN 978-1-59448-561-9 (Riverhead trade pbk.) 1. Green technology—Anecdotes. 2. Energy consumption—Climatic factors. 3. Consumer behavior—Environmental aspects. 4. Sustainable living. I. Title.
TD148.O94 2012
628—dc23

2011028855

PRINTED IN THE UNITED STATES OF AMERICA

10 9 8 7 6 5 4 3

CONTENTS

1

The Conundrum

During the summer of 2010, I gave a talk in Melbourne, Australia, as part of a weeklong state-sponsored series of lectures on climate change. A couple of days before I spoke, a resident asked me what my theme was going to be, and when I began to explain he stopped me. "Forget all that," he said. "Just tell me what to buy." He was willing to believe the world was in peril, but he wished that someone would cut to the solutions. My car's a problem? Tell me what to drive instead. Wrong television set? I'll switch. Kitchen counters not green? I'll replace them with whatever you say.

This is the way most of us think, whether we think we do or not. We're consumers at heart, and our response to difficulties of all kinds usually involves consumption in one form or another: just tell me what to buy. The challenge arises when consumption itself is the issue. How do we truly begin to think about less—less fossil fuel, less carbon, less water, less waste, less habitat destruction, less population stress—when our sense of economic, cultural, and personal well-being is based on more?

The world faces a long and growing list of environmental challenges. Yet most so-called solutions—our strategies for "sustainability"—either are irrelevant or make the real problems worse. We deceive ourselves in myriad ways. The bestselling vehicle in the United States, both today and historically, is the Ford F Series pickup truck; it now comes with an optional "EcoBoost" engine, which gets sixteen miles to the gallon in non-highway driving. In 2009, Toyota released a television commercial in which a Prius—the popular gas-electric hybrid, which gets better mileage than comparable vehicles with ordinary gasoline engines—glides across a colorless landscape. As the car passes, the ground bursts into bloom, happy music plays, children with butterfly wings float above children swaddled in flower petals, clouds come alive, and buildings in the distance glow—as though the car were vacuuming bad things from the world, truly establishing what the announcer describes as "harmony between man, nature, and machine." In 2010, a forward-thinking friend of mine took me for a ride in another hybrid, a Ford Fusion. As he employed such fuel-saving procedures as braking smoothly and avoiding jackrabbit starts, his dashboard fuel gauge filled with images of intertwining green foliage, a symbolic representation of the environmental benefits we were apparently dispensing from the tailpipe as we aimlessly drove around. I've got a college degree, but I felt a twinge of idiotic virtue, as I also do when I leave an especially large pile of cans, bottles, and newspapers at the end of my driveway, for the recycling truck.

One of our favorite green tricks is reframing luxury consumption preferences as gifts to humanity. A new car, a solar-powered swimming-pool heater, a two-hundred-mile-an-hour train that makes intercity travel more pleasant and less expensive, better-tasting tomatoes—these are the sacrifices we're prepared to make for the future of civilization, along with various punitive economic policies aimed at the Chinese. Our capacity for self-deception can be breathtaking. What, for example, was the environmental value of my traveling halfway around the world to give a forty-five-minute speech to a few hundred Australians who, because they'd elected to be present in the first place, were probably predisposed to agree with just about anything I might say? Air travel currently accounts directly for something like 3.5 percent of global energy use and manmade-greenhouse-gas production, and my trip to Melbourne (on which I was accompanied by my golf clubs and my Hertz No. 1 Club Gold card) involved plenty of both. So did the United Nations Climate Change Conference held in Copenhagen in December 2009. It attracted thirty thousand delegates and other registered participants, as well as thousands of reporters, bloggers, observers, supporters, protesters, and others, and most of them had traveled to Denmark from places far away. Forestalling global calamity is a preemptively worthy goal, and, even in the era of Skype and the iPhone, there's no fully satisfactory replacement for face-to-face discussion. Yet, if climatologists, environmentalists, government officials, and others who are earnestly trying to figure out what to do about greenhouse gases can't confer without crossing oceans, what can be expected of those of us who may be less informed, less committed, and less motivated to sacrifice our own convenience?

The point isn't that we're all hypocrites, although of

course we are. (If you're a connoisseur of big carbon footprints, check out the book-tour itineraries and frequent-flier balances of just about any authority on the environment.) The point is that, even when we act with what we believe to be the best of intentions, our efforts are often at cross-purposes with our goals. That's the conundrum.

Captain James Cook sailed from England in August 1768 on the first of the three voyages that would make him famous, and he and his crew reached Australia two years later. Their ship, the *Endeavour*, was powered by wind, but the trip nevertheless required a huge expenditure of natural resources, not least in the form of human fuel and human lives. (Just fifty-six of the ninety-four men who set out from England made it back.) My 2010 trip to Australia, by contrast, was a miracle of efficiency, because my conveyance was the product of many decades of technological innovation. The plane carried almost five times as many passengers and crew members as the *Endeavour*, required no repairs en route, recorded no casualties, made it from New York to Melbourne after less than twenty-four hours in motion, and was just one of hundreds of international voyages arriving in the country that day.

My flight consumed a lot of energy; in fact, my proportional share of the fuel burned during my round trip was greater than the total amount of energy that the average resident of the earth uses, for all purposes, in a year.* But the environmental problem with modern flying is not that our airplanes are wasteful; the problem is that, thanks to the steady application of engineering brilliance, we have eliminated so much waste from long-distance journeys that, nowadays, the main impediment to traveling ten thousand miles for a week's visit is less likely to be the cost of the ticket (in early 2011, as little as six or seven hundred dollars each way between New York and Melbourne, if not "free" through the redemption of miles) than the perceived unpleasantness of spending a whole day watching movies and sleeping in a cushioned, reclining seat.

When people talk about reducing the energy and carbon footprints of air travel, they almost always focus on

---

* This calculation is based on figures in *Sustainable Energy—Without the Hot Air*, by the British physicist David J. C. MacKay.

things like improving the design of engines, wings, and fuselages, and perhaps using computerized control systems to shorten flight paths and eliminate delays. By this point, though, the total potential gain in any of those areas is small. Today's passenger jets are already something like 75 percent more fuel efficient than the jets of the early 1960s, and the physics of flying imposes a low ceiling on further advances. But even if the potential for additional reductions were huge the main effect of any such innovations would be the same as the main effect of all such innovations since the time of Captain Cook: they would make travel easier, cheaper, more convenient, and more attractive than it is already, and would therefore encourage us to do more of it.

Flying from New York to Melbourne in 1958—on a Lockheed Super Constellation, a propeller aircraft—consumed more energy per person than my flight did, but it was "greener" nevertheless. It required stops in San Francisco, Hawaii, Canton Island, Fiji, and Sydney, and it cost each coach passenger something like a quarter of that year's U.S. median family income, each way. If comparably slow and costly flights were the only Australia travel option available today, I and just about everyone who traveled with me would have stayed home: a gain for the environment. The only clearly, unambiguously effective method of reducing the carbon and energy footprints of air travel is to fly less—a behavioral change, not a technological one.

We already understand how to fly less: no scientific breakthrough required. But where's the fun in going nowhere? Travel is exciting and romantic and educational, and the livelihoods of many millions of people all over the world depend on our continuing to do lots of it. So instead of taking steps to cut back personal mobility we talk about aeronautical innovation and improved efficiency, or we assuage our consciences by making trivial contributions to organizations that promise to "offset" the environmental impact of our wandering. Meanwhile, we continue to build and expand airports. (The world's largest building is in Dubai, but it isn't the Burj Khalifa, the skyscraper that's twice as tall as the Empire State Building; it's Terminal 3 at Dubai International Airport.) We also subsidize airlines and aircraft manufacturers, and devise ways to "efficiently" squeeze even more planes into existing air routes, and spend billions on marketing campaigns whose purpose is to make us and others yearn to be in the air and on our way—perhaps on "eco-tours" to environmentally threatened corners of the earth. That's the conundrum.

In 2004, Saul Griffith, a young Australian who was working on his PhD at the Massachusetts Institute of Technology, won a thirty-thousand-dollar award that is given each year to a student who has shown unusual promise as an inventor. Griffith was an obvious candidate. Neil Gershenfeld, one of his professors, described him to me as "an invention engine" and said, "With Saul, you push 'Go' and he spews projects in every imaginable direction." The judging committee was especially impressed by a device that Griffith had created to custom-manufacture low-cost eyeglass lenses, intended for people in impoverished countries. He had conceived of the process after discussing Third World health issues with the education minister of Kenya, and his system had been recognized already by the Harvard Business School and the National Inventors Hall of Fame.

Traditional lens making is a process that involves thousands of costly molds. Oversized lens blanks are cast in plastic, and then technicians grind and polish the blanks to match prescriptions. Griffith told me, "I wanted to make a machine that would negate the need for that entire factory—to let you print the lenses on demand." So he built a desktop device with which a minimally trained operator could turn a fast-hardening liquid into a lens in a few minutes. The machine consisted of a single, universal mold, with an adjustable metal ring—like a tiny springform cake pan—between a pair of flexible membranes, whose degree of convexity or concavity could be controlled with a simple hydraulic system (in the prototype, a pair of horse syringes filled with baby oil). "Literally with only those two inputs—the shape of your boundary condition and the pressure—you can define an infinite number of lenses," Griffith explained. In 2007, he won a $500,000 MacArthur Fellowship—a "genius grant"—and the MacArthur judges cited the eyeglass invention as having "the potential to change the economics of corrective lenses in rural and underserved communities around the world."

But winning prizes ended up being easier than changing the world, and Griffith's lens printer has never found a market. "It turned out that we were solving the wrong problem," he told me. "A lens factory is expensive to build and equip, but once you've got one you can make lenses cheaply, and then you can deliver them anywhere in the world for a dollar or two in postage." In effect, Griffith's invention addressed a problem that had been solved years before, at lower cost, by Chinese labor and global shipping. The real issue with eyeglasses in the developing world isn't making lenses, he said; it's testing eyes and writing prescriptions for people who have little or no access to medical care—a matter of politics and economics rather than technology.

That experience has deeply influenced Griffith's thinking on the environmental subjects that consume him now. He is still an invention engine; his many recent projects include an electricity-assisted cargo-carrying tricycle, an inexpensive form of insulation inspired by origami, and an unconventional method of generating power with wind. But unlike many engineers and scientists who share his concerns, he told me, he is "not really a techno-optimist"—someone who believes that the world's energy and climate problems will ultimately yield to a redoubled application of human technological ingenuity: an ecological Manhattan Project, a green Apollo program. He has come to believe that the world's most urgent environmental need is not for some miraculous-seeming scientific breakthrough but for a vast, unprecedented transformation of human behavior, a revolution in our relationship with energy and consumption. In a presentation he made in 2009, he said that he intended to give his son (who, at that point, was still a few months from being born) a Mont Blanc pen and a Rolex watch—his metaphors for what he referred to as "heirloom technology," or high-quality, low-energy possessions that are meant to be used over an entire lifetime rather than quickly discarded and replaced.

Even if we believe that Griffith is right, how many of us would freely choose to transform our lives to the extent that he intends? And, whether we would or not, what political, economic, or moral force would be strong enough to bring enough of us into line to make a difference? That's the conundrum.

Beginning in late 2008, something happened that not even the most optimistic environmentalists and climatologists had been expecting: the world's energy and carbon footprints shrank, and by significant amounts. But those changes didn't occur because the human race had suddenly attained environmental enlightenment. They occurred because the price of oil had risen to record levels and the global economy had tanked. The explanation is easy to understand: the world's main emitter of manmade greenhouse gases has always been prosperity, and when times are bad consumers respond by consuming less. Closed factories don't burn coal, and when people lose their jobs, or become worried about losing their jobs, they behave like model members of the Natural Resources Defense Council. They drive less, turn down their furnace, shut off lights in empty rooms, stop heating their swimming pool, make do without air-conditioning, decide not to travel to Chicago for Thanksgiving, and make fewer impulsive purchases. Even wealthy people hunker down, and the resulting interruption in economic growth slows the pace at which we worsen a long list of environmental troubles. The global recession that began in 2008 spread pain through much of the world, but it also put time back on the carbon clock.

No rational person would advocate economic collapse as a climate-and-energy strategy. From an environmental perspective, though, there are things to be said for recessions. The challenge is to find a non-catastrophic way to accomplish the equivalent—to harvest the benefits of falling consumption without inflicting the human suffering. That's an especially tough problem because, as recent experience proves, when times are tough even the world's wealthiest nations focus their efforts not on perpetuating serendipitous reductions in energy use and carbon output but on making them go away—by cutting energy taxes, bailing out bankrupt manufacturers, reducing the cost of borrowing, encouraging sprawl with incentives to builders and home buyers, weakening environmental controls, and opening fragile ecosystems to resource exploitation.

We may believe that we care about the world's deepening environmental challenges and are merely waiting for scientists, environmentalists, politicians, and others to come to their senses and implement effective solutions on our behalf. In truth, though, we already know what we need to do, and we have for a long time. We just don't like the answers. That's the conundrum.

2

Setting Things on Fire

We're richer than kings. We live better than William I, who conquered England a millennium ago but, for all his wealth and power, didn't have a flush toilet. An army of ruthless Normans and the blessing of Pope Alexander II? Yes, he had those things. Antibiotics and a riding lawn mower? No. Although William was stupefyingly wealthy by the standards of his time and place, he seems starkly impoverished by the standards of ours. No blow-dryer, no iPad, no seedless grapes. How did he get by?

History books are filled with the names of supposedly loaded people who, upon closer inspection, turn out to have been destitute in comparison with us. Alexander the Great wasn't an Amazon Prime member. Czar

Nicholas II lacked a dishwasher with a dedicated flatware tray. Poor little John D. Rockefeller: all the oil in the world couldn't buy him a Lincoln Navigator.

Our comforts seem so familiar to us that we seldom pause to be astonished by them. In fact, despite our riches, we're more likely to dwell on what we don't have than on what we do. You'd think that not having bubonic plague would be enough to put most of us in a cheerful mood—but, no, we want a hot tub, too. There's a Web site called *The Coveteur* (www.thecoveteur.com), whose purpose is to inspire acquisitive longing by displaying photographs of other people's possessions: to create a feeling of deprivation where none had existed, and then offer links to make it go away.

Our seemingly unquenchable sense of dissatisfaction has served us well in innumerable ways: it's the engine of civilization. But it's also a source of trouble—increasingly so.

When we do ponder our luxuries, we don't think very hard about where they come from. Mostly, we attribute them to the inventiveness of our species—a reasonable conclusion, since you don't see other primates, or even dolphins, planning trips to Mars. The British writer Matt Ridley, in *The Rational Optimist*, assigns the main credit to the apparently unique ability of humans to innovate collaboratively, enabling all of us to enjoy benefits that none of us could have executed or even conceived alone—an especially productive form of co-operation which Ridley describes as ideas having sex.

But the ultimate source of our prosperity is simpler, and more fundamental than intelligence in any configuration: it's combustion. Our history is the chronicle of our ascent up what the naturalist Loren Eiseley called "the heat ladder," and modern life is mainly the product of our steadily growing talent for usefully setting things on fire—wood, coal, oil, natural gas. The Bronze Age could more accurately be called the Fire Age, because without combustion virtually all metals would remain where they were when our world began, as would almost all the other extracted treasure on which modern life depends. If we had to get by on muscles, brains, and working together, we'd still be chasing zebras or huddling in caves. What economists call "income" and "economic growth" are almost entirely by-products of burning and burning-enabled resource exploitation; so are virtually all the incalculable blessings of modern life, including the luxury of being able to say that our most important possessions are not the material ones.

Heat is energy, and energy is leverage. Our growing cleverness at setting things on fire has hugely extended our strength and the reach of our consumption, as well as magnifying our ambition to consume more. A few barrels of oil or tons of coal, ignited, yield the energy equivalent of a lifetime's worth of a healthy man's unassisted physical labor. Even when we do what we think of as working with our "hands," we are heavily dependent on burning, since without combustion we would have no tools or raw materials or sheltered spaces in which to use them. We sense our dependency every time a winter storm knocks down the power lines, but such glimpses are fleeting, and they are only minimally illuminating because they don't convey the scale of our need. After all, we can still use our cars to recharge our cell phones.

Without truly thinking about it, we burn tremendous amounts of fuel. The BP oil spill in the Gulf of Mexico in the spring of 2010 was a major news story for months, so much so that some people were given actual nightmares by televised images of the dark plume gushing from the seafloor. (A Web site called IfItWereMyHome.com created an interactive page that enabled anxious users to depress themselves further by superimposing a spill-sized gray blob anywhere on a Google map of the world—making it possible to just about smother Virginia, for example.) But the true environmental catastrophe wasn't the oil that went into the water; it was the oil we continued to use exactly as we intended. The average flow rate from the BP wellhead was fifty-five thousand barrels a day. That's a huge amount, but Americans use that much every four minutes, and the estimated total BP spill volume—nearly 5 million barrels over eighty-seven days—is consumed by the world every hour and a half. Current average global oil consumption is the equivalent of more than 1,400 BP spills gushing nonstop, around the clock, every day of the year. And that's just the oil.*

Saul Griffith, the Australian engineer whose experience with cheap eyeglasses I described in the first chapter, once gave a lecture during which he hung an IV bag filled with oil on a stand beside the podium and allowed the oil to drip at approximately the rate that the average American consumes it (about a pint per hour). He has

---

* A 2010 *Onion* headline: "Millions of Barrels of Oil Safely Reach Port in Major Environmental Catastrophe."

also asked audiences to reify their energy consumption by imagining themselves leaving home each morning wearing a backpack containing the fossil fuels their lives will require that day: for the average American, roughly twenty-two pounds of coal, 180 cubic feet of natural gas, and almost three gallons of oil. Almost nothing in human history comes close to that extravagance. Vaclav Smil, in his 2003 book *Energy at the Crossroads,* writes that the (mostly fossil-fuel-derived) power under the direct control of the average American household, including its vehicles, "would have been available only to a Roman *latifundia* owner of about 6,000 strong slaves, or to a nineteenth-century landlord employing 3,000 workers and 400 big draft horses"* [p. 58]. By burning the world, we have made ourselves rich and healthy and comfortable on a scale that few of our ancestors could have imagined. Yet we seldom acknowledge the underlying source of our astounding prosperity—which Smil described succinctly in an e-mail in 2011: "Moneys are (however inadequate) a

---

* Also highly recommended is Smil's *Energy Myths and Realities,* which was published in 2010.

measure of energy flows (the only real currency in the biosphere)." Conventional economists tend to treat affluence as more nearly a cause than a consequence of energy use. In fact, though, any economy, no matter how "efficient," is a fuel-consuming machine.

3

Fossil Fuels as Credit Card

Even though our debt to combustion is huge, the true nature and scale of our need are easy for even experts to minimize. A popular concept among some economists and environmentalists is "decoupling," which suggests that the growing efficiency of our machines has weakened the link between energy and prosperity—as indicated by the fact that in the United States expenditures on energy production constitute a small and falling fraction of gross domestic product, and by the observation that employed Americans nowadays are more likely to spend their weekdays standing behind counters or sitting at desks than stoking furnaces in smoke-spewing factories. A similar concept is "decarbonization," which states that—because the fuels we burn today have a lower carbon content than fuels we burned in the past, and because we burn them more cleanly and more productively—every dollar we spend represents a shrinking quantity of greenhouse gas.

These sound like environmentally invaluable trends, since they suggest that we are reducing a long list of undesirable impacts by cleverly finding ways to get more for less. And both concepts are spoken of not only as reasons for environmental optimism but also as tools for negotiating climate agreements with countries that, like China, oppose constraints on their overall energy use and greenhouse emissions. (The idea is that such countries, instead of cutting back in absolute terms, thereby dampening their economic growth, would agree to make every dollar's worth of energy or carbon go further in terms of GDP—in their view, a far more appealing proposition, since they're doing it anyway.) But both concepts suggest, perplexingly, that the world's energy and carbon challenges have been solving themselves for decades, since decoupling and decarbonization are nothing new: they are mainly reflections of the steady growth in productivity which has almost always characterized civilization's relationship with fuel-burning machines. (The main part of the Industrial Revolution was inaugurated by a major advance in decoupling and decarbonization: James Watt's invention of a more energy-efficient steam engine.) The environmental problem with such advances is that the productivity gains have almost always been reinvested in additional production: as we've gotten better at making things, we've made more things. That has made us astonishingly richer, healthier, more comfortable, and more numerous, but it has also incurred a mounting environmental cost. Believing that ongoing decarbonization is an irresistible force for environmental good, furthermore, depends on a certainty that the world is on the verge of making a major, historically inevitable transition to lower-carbon power—say, to nuclear fission, the next rung on the heat ladder—rather than becoming less picky about what we choose to burn: tar sands, shale gas, high-sulfur coal, trash, forests. It also depends on a belief that decarbonization has the force of natural law, and that our energy use is irresistibly evolving toward a climate solution.

A more modern difficulty with the concepts of decoupling and decarbonization is that, as they're often applied, they don't account for energy use and carbon emissions that have been not eliminated but merely exported out of the region under study—say, from California to a factory in China. Indeed, much of what's described as the greening of various parts of the U.S. economy may merely be a consequence of the fact that we Americans now physically manufacture, within our borders, a smaller percentage of the stuff we buy and throw away. (The carbon and energy footprints of our imports may now greatly exceed those of our exports—a new kind of trade imbalance.) Shifting more manufacturing to plants in Asia and elsewhere has made America's environmental accounts look cleaner than they would have otherwise, but it hasn't improved the environmental outlook of the world. Indeed, it has made that outlook worse, by making destructive processes and products cheaper and by moving energy-hungry industries out of the purview of the U.S. Environmental Protection Agency and into that of governments that allow the burning of nastier fuels and impose fewer environmental controls. It has also weakened Americans' sense, never strong, of the consequences of consumption, since many of the worst direct impacts now occur elsewhere—for example, in the choking, mustard-colored skies of southern China and the fouled waterways and poisoned slums of India.

And there's a more fundamental problem, described by the Danish researcher Jørgen S. Nørgård, who has called energy decoupling "largely a statistical delusion." Nørgård—in an essay in *Energy Efficiency and Sustainable Consumption*, which was edited by Horace Herring and Steve Sorrell and published in 2009—writes:

> Every economic activity requires some energy consumption and all energy consumption is rooted in some economic activity, be it on the consumer or producer side. The observation that the two parameters can grow at different rates, which over history is quite normal, does not imply any decoupling. Whether a car is running in first or second gear, there is still a coupling between the engine and the wheels, and speeding up the engine will speed up the car. Unfortunately, the notion of a decoupling has served as peacemaker between environmentalists and growth-oriented politicians by conveniently exempting economic growth of any responsibility for environmental problems.

To say that energy's economic role is shrinking is a little like saying, "I have sixteen great-great-grandparents, eight great-grandparents, four grandparents, and two parents—the world's population must be imploding." The cost of producing energy may directly account for a small percentage of the U.S. economy, but its falling share of GDP has actually made it more important, not less, since every kilowatt supports an expanding proportion of our well-being. Switching off America's entire electricity-generating capacity in 1882 would have caused four hundred New York City lightbulbs to go out; switching off America's entire electricity-generating capacity today would cause a national (and therefore global) catastrophe beyond imagining.

Asserting that energy plays a shrinking economic role, furthermore, is the rough equivalent of asserting that, during the buildup to the recent collapse of the American housing bubble, the ability to make monthly mortgage payments played a shrinking role in Americans' fitness to own homes. The eagerness of lenders to finance huge purchases by people who couldn't afford them made expensive houses seem cheaper, but it did nothing to change the *actual* economics of home ownership. All it did was to "decouple" real wealth and purchasing power—and, by increasing the leverage of unqualified buyers, it magnified the eventual disaster. Energy leverage works in a similar way. The growing cleverness of the human race at turning combustion into affluence doesn't make us *less* dependent on fossil fuels. On the contrary.

At any rate, if believers in the inevitable environmental benefits of decoupling and decarbonization are right, we've got little to worry about, since our energy and carbon troubles will continue to shrink of their own accord, as the weight of history presses them into insignificance. So the prudent thing would be to assume, at least for the time being, that they might be wrong.

4

LEED-Certified Landfill

Perhaps most important of all, without combustion we would have minimal access to the great secondary source of our tremendous wealth: the extraction, transformation, and incremental exhaustion of the earth's vast but ultimately finite inventory of exploitable natural resources, including combustible fuels themselves. Our growing skill at generating heat and turning heat into motion and turning motion into power has enabled us to withdraw riches from the earth with such ease and in such quantities that we have little sense of the real roots of human prosperity. Some economists and environmentalists speak of economic "dematerialization"—by which they mean our growing ability, through innumerable engineering advances, to produce more finished goods from less raw material. Yes, we have more possessions than our parents did, this way of thinking goes, but each possession is lighter and smaller, in relation to its function, than the objects it evolved from, and creating it involved less waste. The implication—as with energy and carbon—is that, if we just keep consuming, our environmental problems will gradually shrink to a manageable size, as we continue our gentle glide toward the bottom right-hand corner of the graph.

But thinking this way depends on a willful determination to study small pictures instead of large ones. Making more goods from fewer inputs reduces the material content of each good, but it also makes those same goods cheaper and, ultimately, more plentiful and expendable, and all of that additional buying and selling makes us richer, and as we grow richer we have the means and the will to buy even more. (An ever growing share of China's manufacturing output is consumed locally, rather than exported, now that selling disposable luxuries to us has made the Chinese rich enough to buy disposable luxuries of their own.) One consequence is that we now acquire new possessions in such a torrent that we often can't think what to do with the old ones—like the fully functional but now painfully obsolete-seeming television sets that we happily watched until high-definition flat-screens came along. (My own lengthy delay in buying a fancy new TV was caused mainly by my inability to figure out how to get rid of my previously beloved Trinitron, which was too huge and heavy for even two people to carry out the door.) A thriving niche in the U.S. economy in recent years has been the self-storage industry, which rents supplemental space to people whose attics, garages, and super-sized houses are overflowing. (In 2010, the American self-storage industry had revenues of about $20 billion—more than ten times Facebook's.) A friend who rents a self-storage unit in my town told me he signed up because his house was becoming cluttered but he hated the idea of sending all his excess stuff to the local landfill. Yet self-storage, for most of its customers, is really just landfill with a roof and a lockable door. And that's true even if the facility they use—like an ezStorage location in Elkridge, Maryland, in 2011—has been certified "Gold" by the U.S. Green Building Council's increasingly popular eco-rating system, Leadership in Energy and Environmental Design, known as LEED.

Not that many of us hesitate to send unwanted possessions to actual landfills. We Americans routinely discard so much that our country now spends more on plastic garbage bags than almost half the world's countries spend on everything. Of course, garbage bags are prime examples of dematerialization, since each modern one contains a fraction of the plastic that the Naugahyde-gauge Glad bags of my youth did, back in the sixties. But has reducing the cost and increasing the availability of garbage bags made us less profligate? On my desk I have an old beer can, from the 1940s, that once contained twelve ounces of Hampden "mild but sturdy" ale. The empty can (which I found inside a wall in my house during a renovation project) weighs seventy-nine grams, or five and a half times as much as a modern twelve-ounce beverage can made of aluminum. That modern can represents an impressive feat of dematerialization. But has the slimming of our disposable containers caused the per capita human waste stream to shrink? Or has it merely enabled and encouraged us to become still more reckless in our consumption?

5

Problems Innovate, Too

We tend to think, as we ponder strategies for overcoming various environmental difficulties, that technological innovation is a purely benign force. But problems innovate, too—and, usually, they have better funding. Engineering breakthroughs enable machines to do more work with less fossil fuel, but they also do things like making it possible to extract fossil fuel that used to be inaccessible or unknown and to manufacture new and highly desirable products that, in addition to swelling our contentment as consumers, further increase our demand for fossil fuel. We clearly need brilliant people to develop and exploit cleaner energy sources, but, meanwhile, their efforts are being undermined by equally brilliant people who are working, even harder—and with bigger budgets and higher salaries—to expand our inventory of fossil fuels and invent new things to do with them. On June 10, 2011, the editorial writers of the *Wall Street Journal*—who have long been among the nation's most vocal proponents of global inaction on climate, renewable energy, and a long list of other environmental issues—wrote, "The great irony of recent years is that governments have thrown hundreds of billions of dollars at wind, solar, ethanol and other alternative fuels, yet the major breakthroughs have taken place in the traditional oil and natural gas business. Hydraulic fracturing in shale, horizontal drilling and new seismic techniques are only the best known examples." And this is true—although it's not an environmental positive. The biggest recent blow to the development of renewable energy has been the spread of hydrofracking, which has pushed down the cost of natural gas—the "good" fossil fuel—as well as significantly enlarging the world's supply of recoverable oil. In almost any innovation race, carbon-based fuels are going to win. (For more on hydrofracking and natural gas, see Chapter 23.)

Scientists often cite the elimination, worldwide, of almost all the industrial uses of chlorofluorocarbons

(CFCs)—which are ozone-destroying chemical compounds that were once used widely as refrigerants (in things like refrigerators and air conditioners) and propellants (in things like hair spray and deodorant)—as an example of the ability of scientists to deal rapidly with an environmental crisis when backed by sufficient political support and funding. But they seldom talk about the role that scientists played in creating the ozone problem in the first place. After all, it was scientists who invented CFCs, figured out how to exploit them commercially, and helped to rapidly develop a lengthening list of profitable uses for them.

The point isn't that science is evil, but merely that innovation seldom behaves exactly the way we think we want it to. Almost all the serious environmental problems we face now are the direct or indirect consequences of what seemed, originally, like awfully good ideas. Relying on technology to solve those problems means having faith in our ability to eliminate or contain the inevitable unintended consequences—a big gamble, if history is an indication. (A leader in the development of CFCs was Thomas Midgley, Jr., a chemist and mechanical engineer, who could probably be considered the patron saint of unintended consequences, since he was also instrumental in the development of the gasoline additive tetraethyl lead. Midgley, according to the environmental historian J. R. McNeill, "had more impact on the atmosphere than any other single organism in earth history." He was also extremely unlucky. His work with lead gave him lead poisoning, from which he took a year to recover, and in 1940 he contracted a severe case of polio. He invented a hoist system to enable polio victims to lift themselves from bed, and died in 1944, of asphyxiation, after becoming entangled in the ropes.) The reduced use of CFCs also represents an instructive example of the power of the unforeseen. In refrigerators, air conditioners, vending machines, and other compressor-based cooling equipment, CFCs have largely been replaced by fluorocarbons (FCs) and hydrofluorocarbons (HFCs), which are compounds that have similar properties but are not harmful to the ozone layer. But FCs and HFCs are potent greenhouse gases, as CFCs also are—a problem that didn't become obvious until later. According to an article in *Scientific American* in 2011, "the most commonly used gases, when released, warm the atmosphere at least 1,000 times more than carbon dioxide does, molecule for molecule." The use of such gases will eventually have to be restricted or eliminated for that reason, but there's no guarantee that whatever we replace them with won't cause trouble, too, in some way we didn't anticipate. Meanwhile, the world's rapidly growing inventory of refrigeration equipment constitutes a vast, expanding reservoir of climate-altering compounds, which will eventually have to be recovered and neutralized or permanently sequestered. How good are we likely to be at doing that?

6

The Greenest Community in the United States

The greenest community in the United States isn't Portland, Oregon, or Boulder, Colorado, but New York City. To many people, including many New Yorkers, that idea seems perversely contrarian, but the evidence is straightforward. New Yorkers, individually, use less energy in all forms than any other Americans, and they have the smallest carbon footprints (less than 30 percent of the U.S. average). Not coincidentally, they are also the country's only truly significant users of public transportation. The New York metropolitan area accounts for nearly a third of all the public-transit passenger miles traveled in the United States, and the city itself contains half of all the country's subway stops. In addition, New

Yorkers are the last large U.S. population for whom walking is still a primary form of transportation. (In suburbia, when you spot people on foot they're almost always either moving between vehicles and buildings or trying to lose weight.) The United States today is so dependent on automobiles that the average household owns more than two vehicles, and more than a third of all households own three. (In South Dakota, nearly 13 percent of households own five or more.) In New York City, by contrast, 54 percent of all households don't own even one car—in Manhattan, the figure is 77 percent—and most of the families that do own a car don't drive it the way other Americans do. (Manhattanites use their cars mainly to make periodic escapes from Manhattan; residents of Queens, Brooklyn, and the Bronx use theirs mainly on weekends; residents of Staten Island use theirs more like average American suburbanites, though not as much.)

The fundamental reason for New York's leadership in all these categories is the very thing that, to most Americans, makes the city look like an ecological nightmare: its extreme compactness. New York is, by far, the most densely populated U.S. city, with more than twenty-seven thousand residents per square mile; Manhattan (which is the smallest of the city's five boroughs in land area) is even denser, with sixty-seven thousand people per square mile, or eight hundred times the average density of the country as a whole. Squeezing people close together may not look green—where are their solar panels, carbon-sequestering trees, and backyard compost heaps?—but it actually reduces environmental impact, because it dramatically shrinks car ownership, makes efficient public transit possible, constrains energy use in all categories, and forces most residents to live in apartment buildings, which are among the world's most efficient residential structures. High-density living also sharply limits residents' opportunities for personal consumption and waste. New Yorkers don't have lawns, sprinkler systems, swimming pools, or rooms they seldom set foot in, and—because living space is tight and expensive—they don't accumulate large inventories of energy-sucking household appliances. The city's visitors (and New Yorkers themselves) often complain about garbage on the streets, but New Yorkers, individually, generate less solid waste than other Americans do: with less room for acquisitions, they acquire fewer things and, therefore, throw fewer things away. (In suburbia, garages nowadays mostly contain not cars but surplus stuff, including the unused recreational equipment and outdoor furniture of people who seldom do anything in their yards other than working on them or watching other people work on them.) New Yorkers also use less water than other Americans do, because, with no lawns or swimming pools, they have fewer opportunities to use it.

Intelligently increasing population density—shortening the distance between people, and between people and their destinations—is the key to reducing a long list of negative environmental impacts in mobile, affluent populations. Worldwide, the prosperous communities that use the least energy and do the least damage to the environment are those in which, as in New York, living spaces are small, residential and commercial uses are interspersed, and per capita car use is very low—places like Hong Kong, Tokyo, and the older sections of European capitals (which have the significant environmental advantage of having been laid out before the invention of the automobile). Hong Kong contains one of the densest concentrations of people and wealth on the planet, yet the average resident uses only about a third as much energy as the average resident of the United States. Intelligent density is the reason.

Such arguments run counter to much conventional thinking about low-impact living. In 2007, *Forbes* assessed the environmental profiles of the fifty U.S. states and picked Vermont as the greenest. It's true that Vermont has an abundance of trees, farms, compost, and environmentally aware citizens, and it has no crowded expressways or big, dirty cities. (The population of the state's largest city, Burlington, is just under forty thousand.) Vermont also ranks high in almost all the categories on which *Forbes* based its analysis, such as the proportion of LEED-certified buildings and the implementation of public policies that encourage energy efficiency.

But *Forbes*'s analysis was misconceived, because Vermont actually sets a poor environmental example. In the categories that matter the most, Vermont ranks very low. It has no significant public transit (other than its school bus routes), and, because its population is so dispersed, it's one of the most heavily automobile-dependent states in the country. A typical Vermonter consumes 545 gallons of gasoline per year—almost a hundred gallons more than the average American and six times as much as the average Manhattanite. (Among the fifty states, Vermont ranks fourth highest in per capita expenditure on gasoline, while New York State, entirely because of New York City, ranks last.) Spreading people thinly across the countryside, Vermont style, may make them feel green—and look green to *Forbes*—but it actually increases the damage they do to the environment while also making that damage harder to detect and undo. An easy way to see that is to imagine dismantling New York City and dispersing its 8.4 million residents across the countryside at the population density of Vermont. To do that, you'd need living space equal to the combined land area of Maine, New Hampshire, Massachusetts, Connecticut, Rhode Island, Delaware, New Jersey, Maryland, Virginia, and Vermont itself—and then you'd have to find places to put all the people you were displacing, as well as providing them with the vast, redundant infrastructure of suburban sprawl: the roads and sewers and power lines and gas stations and hospitals and shopping malls and outlet stores and schools and parking lots and everything else.

Conversely, the combined population of Alaska and Montana—two states that, between them, contain a fifth of all the land area of the United States—is only slightly larger than the population of Manhattan. And Wyoming and North Dakota, which are No. 1 and No. 4 among the fifty states in terms of highest energy consumption per resident, have, between them, a population smaller than the Bronx. These demographic facts suggest the possibility of a troublesome political divide, in terms of legislative approaches to energy and climate issues: the four high-consumption states just mentioned are represented by a total of eight U.S. senators, while Manhattan and the Bronx, with the same population but a fraction of the carbon footprint, share less than a third of one.

7

Learning from Manhattan

My purpose in extolling the environmental benefits of population density is not to argue that the residents of Manhattan, Hong Kong, and central Paris are morally superior to people who live in Vermont or Wyoming, or to suggest that everyone should pack up and move to Tokyo. There are many downsides to density, including the fact that squeezing people and their destinations close together makes diseases, wars, and natural disasters more efficient, too. The earthquake and tsunami that devastated parts of Japan in early 2011 would have inflicted far fewer casualties and far less property damage if the main impact had been in a more sparsely populated area. Nor am I unaware of the many unpleasant lifestyle aspects of living in dense urban cores, including the noise, bad smells, smog, vehicle exhaust, crime, tiny apartments, lousy schools, depressing playgrounds, scarcity of vegetation, and ever-present crush of humanity. My wife and I and our one-year-old daughter fled Manhattan in 1985 for all of those reasons and any number of others, most of which can be reduced to an ultimately irresistible desire to raise our kids the way each of us had been raised: in a big house with a big yard and a driveway full of cars. And neither of us has ever seriously regretted the move (although both our children now live in Manhattan).

But none of that changes the fact that, in terms of personal impact, my family's move from city to country was an ecological disaster. Our energy use soared, our carbon footprint swelled by many sizes, our inventory of furniture, appliances, and seldom-used possessions rapidly expanded to fill the far greater volume of our new home, and we made a discovery common to all former urbanites: when you move from the city to the country what you really do is move into a car—and move your children into car seats—because for virtually all purposes driving is the only way to get around. (One of the first things my wife and I did in our new home was to gain ten pounds each.) Our lives at first seemed "greener" than they had when we were city dwellers, because our house, which is 220 years old, is across a dirt road from a nature preserve, and we see wild animals all the time. But by every objective measure the list of our environmental sins has grown impressively. Simply as a result of our moving ninety miles into the hills, our energy and carbon footprints climbed from the bottom of the U.S. range to very near the top.

What all this means is that, in spite of what intuition often tells us, when it comes to reducing a long list of ominous impacts, densely populated, mixed-use urban centers are far better environmental role models than the New England countryside: dense cities scale, and leafy exurban paradises don't. The main lessons can be stated simply. In order to soften our main environmental impacts, we need to find ways, globally, to *live smaller, live closer*, and *drive less.*

Living smaller means occupying smaller living spaces, as city dwellers do by necessity. The size of the average American house has doubled since 1950, even though the average American household has shrunk. (Meanwhile, the size of the average New York City apartment, though small to begin with, has gotten smaller.) Bigger living spaces mean more raw construction materials, more furniture, more appliances, more heating, more cooling, and, in every application, more energy forever. My wife and I are by no means alone among American homeowners in having a living room we seldom enter except to fluff up the sofa cushions and vacuum the rug. Bigger living spaces also inevitably mean more cars, more spacious yards, longer streets, bigger highways, more parking lots, more civic infrastructure, and all the other accoutrements of suburban sprawl.

Living closer means reducing the distance between ourselves and everywhere we go. This is an especially tough one for us Americans, because in many ways our national history, beginning with the Pilgrims, is the record of our determination to put space between ourselves and others. Sprawl has been our manifest destiny virtually from the beginning, and we've embedded it not only in our psyches and national infrastructure but also in our legal system, through land-use regulations that stipulate minimum separations rather than maximums and whose guiding principle often seems to be a conviction that every citizen has an inalienable right to park. And distance, in a paradoxical way, is self-magnifying: when my wife and I lived in New York, I thought nothing of walking a couple of miles doing ordinary errands, because on a busy city street lined with closely spaced stores there are always interesting things to look at and think about, and because the alternative modes of transportation are usually unappealing; in my little Connecticut town, by contrast, people often drive the 150 yards between the grocery store and the bookstore, because a paucity of intermediate destinations makes that modest distance seem far longer than it is.

Driving less means weakening our profound dependence on cars—another exasperating challenge for Americans, because, of all the innumerable luxuries we take for granted, the one that comes the closest to defining us, in our own eyes as well as the world's, is the automobile. A still greater difficulty is that, because the United States is so young and North America is so spacious, we Americans have spread ourselves out to such an extent that no other form of transportation can readily or conveniently serve us. People tend to think of cars as an "urban" phenomenon, but, in fact, the Americans who drive the most are the ones who live the farthest from central cities, because they have the fewest other options and because their daily destinations are separated by the longest distances. This has always been true: two-thirds of the first million customers of the Model T were people who lived not in cities but on farms or in small towns. Today, most of us live so far from one another and from our daily destinations that going even twenty-four hours without our own wheels can feel like a major deprivation. Every time the price of gasoline spikes, Americans talk earnestly about public transit, but the truth is that in most of the United States efficient transit is impossible. And the same is true in Canada and Australia, which, with the United States, constitute the global Big Three of automobile dependence and per capita energy use and carbon footprint.

The discouraging fact is that, below a fairly high density level of people and destinations, transit simply can't work. In the late 1970s, Jeffrey Zupan and Boris Pushkarev of New York's Regional Planning Association determined that the density threshold for efficient transit is around seven households per acre, or something like double the density of a typical American suburb. And the real environmental benefits of transit don't begin to emerge until densities reach much higher levels—levels at which even buses and trains begin to seem inconvenient, and people simply walk. Phoenix, Arizona, has a nice, modern light-rail system, built at huge public expense, but passenger loads are small, operating deficits are huge, and per-passenger energy consumption is higher than that of driving. That system will never make sense, economically or environmentally. The reason is that the city's residents have sprawled across so much former desert that no one who lives within the metropolitan area will ever be able to rely on transit exclusively, or even mainly, no matter what gasoline is selling for. Greater Phoenix has a population roughly twice that of Manhattan, yet it covers two hundred times as much land. And, as Phoenix keeps spreading, the problem grows steadily worse.

8

Unconsciously Green

Population density is the key to all three of these basic lessons. In fact, merely by *living closer* to one another and to the places where we work, shop, and play, we also, inevitably, *live smaller* and *drive less.* And one of the most appealing facts about higher density, as an environmental strategy, is that its benefits are structural: city dwellers use less energy not because their consciousness is "greener" but because the way they live makes it hard for them to use more, even if they think climate change is a left-wing crock. New York City residents are usually as surprised as anyone when they learn they have the smallest energy and carbon footprints in the United States. That means their low-energy behavior is mostly unconscious—and that's a good thing because unconscious behaviors don't have to be enforced.

What we do about all that is another matter. We're obviously not going to bulldoze metropolitan Phoenix and Atlanta and replace them with something more compact. And even when we do attempt to do the right thing we often choose plans that make the underlying problems worse. New transit in the United States is usually built at the outer edges of existing systems, extending them farther into what used to be sparsely settled countryside. When transit is used in that way, it encourages sprawl, by making it easier for people to live at still greater distances from the places where they work and shop, and by supporting residential growth in places where all other daily travel is necessarily by automobile. It would make more sense, environmentally, to add transit in areas that are dense to begin with, thereby making it easier for urban dwellers to live without cars and encouraging others to do the same. But doing that is harder, both logistically and politically, than using transit to encourage suburban expansion.

We are also hobbled by outdated ideas. A few years ago, a developer sought to build a transit-and-walking-oriented apartment building in downtown Seattle. The building—which was intentionally designed with a limited number of parking spaces—was meant to increase the number of resident pedestrians and transit users in an area that was already quite dense, in comparison with the rest of the United States. The new residents' daily presence would bring more customers to nearby merchants and, by doing so, further enhance the feasibility of living downtown without a car: a complex of positive feedback loops. But the project was turned down, in part because of objections raised by old-style environmentalists, who have often treated almost any kind of urban development (other than the creation of parks, community gardens, and farmers' markets) as environmental outrages. One of the weapons used by the opponents was the city's zoning ordinance, which called for more parking spaces than the plan was designed to provide. Their argument was that the building's residents would inevitably own more cars than the developer intended and would park them in the streets, worsening congestion. Conflicts like this one arise all the time, and often in surprising contexts. We may talk about driving less, but most of us are still worried about where we (or our neighbors) are going to park. Meanwhile, much of the rest of the world seems increasingly determined to follow our dead-end example. General Motors, nowadays, should probably be considered a Chinese car manufacturer, since more of the company's automobile sales occur in China than in the United States, with the result that you're far less likely to spot a Buick Park Avenue on Park Avenue than on almost any street in Beijing.

Even when architects and planners set out to build pedestrian-friendly environments, they often miss the point. They place severe restrictions on building height, or they limit the number of stores and other walkable destinations in residential areas, or they set buildings back from busy streets and sidewalks, or they create oversized plazas that are intended to entice pedestrians but actually resist human penetration in the same way that cavernous living rooms do. (People who are on foot aren't attracted to vacuous open areas the size of parking lots; they're attracted to interesting, miscellaneous, tightly spaced, easily accessible places they want to go, as in central London or, for that matter, inside a large American suburban shopping mall.*) Battery Park

---

*Except for the fact that people have to drive to get to them, suburban shopping malls epitomize the basic principles of the density-oriented urban-planning movements known as Smart

City—a planned community built on ninety-two acres of reclaimed land at the southwestern end of Manhattan—was the joint creation, beginning in the early 1980s, of some of the country's most thoughtful urban planners and designers. Its population density is well above that of the city as a whole (though below that of the rest of Manhattan), and there's a lovely park along the Hudson. Yet most of Battery Park City is a pedestrian dead zone, especially after dark. There are plenty of residents, but unless you're on your way to or from work, there aren't many pedestrian destinations. It's more like a vertical suburb than a vibrant urban core.

Despite all this, it is within our power to manage new development in ways that reduce or contain the overall problem, rather than steadily making it worse. In the United States, our ongoing economic struggle has been environmentally beneficial, because it has dramatically

---

Growth and New Urbanism: the walkways are wide, the shops are tightly spaced, the storefronts come right up to the walkways, and customers go from store to store on foot. If people lived in apartments upstairs, and got rid of their cars, malls would be indoor Manhattans. How about building assisted-living facilities on top of existing shopping malls?

slowed—and in some areas actually reversed—the helter-skelter suburban growth that we have always treated as our birthright. But are we now thinking harder about how to manage growth, or merely catching our breath before beginning another building spree?

9

Sierra Club or Manhattan Club?

For years, the Sierra Club has run a thoughtful campaign, called Challenge to Sprawl, which encourages denser, reduced-impact residential and commercial development and calls for the adoption of more enlightened land-use regulations—a collection of planning ideas usually referred to as Smart Growth. In a perverse way, though, many other activities of the Sierra Club and other American environmental organizations have directly encouraged the opposite, by demonizing cities and inculcating the notion that authentic living is possible only through direct, personal interaction with the natural world. American environmentalists and others usually refer to low-density residential development as *urban* sprawl—a dangerous misnomer, since sprawling subdivisions are actually the antithesis of cities. This same bias causes us to think "sprawl" when we see five houses on one acre, but "green" when we see one house on a hundred, as with a cabin at the end of a dirt road in Montana or a farmhouse framed by fall foliage in Vermont.

When my wife and I left New York with our young daughter, we were driven partly by a desire to possess for ourselves the thing that conventional environmentalists have always extolled: we wanted our own piece of green, and we wanted it right at our back door. But every such exodus increases the outward pressure on the boundaries of development: others follow our appealing example, and when they do we feel that our own personal connection to nature has been weakened, so we move a little farther along. As we do that, we tell ourselves that we are escaping sprawl, but what we're really doing is spreading it.

A long-standing but unexamined tenet of conventional thinking about the environment has been that humans must personally experience unspoiled places in order to value them. This is, among other things, the motivating principle behind the often-repeated notion that children can't be expected to appreciate and protect wonders they've seen only on TV or YouTube. But this idea, though immensely powerful and alluring, doesn't actually align very well with our history as ecosystem destroyers. The tools that have been our principal amplifiers of environmental desolation—trains, cars, airplanes—are also the tools that have done the most to enlarge our opportunities for intimacy with nature. (Rapid growth in attendance at national parks followed rapid growth in ownership of automobiles.) The world's human population is roughly 7 billion, and it will reach at least 9 billion by midcentury or so. How much continued intimacy with people is nature likely to be able to stand?

A more productive mindset might be to turn the conventional wisdom inside out, by focusing on the power of well-conceived urban centers to contain and moderate human impacts on relatively unspoiled areas and their resources—a Manhattan Club, in effect, rather than a Sierra Club. The way to protect the wild places is to concentrate human settlements and make people want to live in them, not to subtly encourage sprawl by treating cities as soul destroyers. That doesn't mean that we all need to cower inside our skyscrapers, watching National Geographic documentaries on TV. But it does suggest that we need to manage human interaction with the natural world differently from the way we have traditionally done.

10

Sierra Club or AARP?

Large numbers of Americans, during significant portions of their lives, actually do live smaller, live closer, and drive less. College students usually live even more compactly than Manhattanites or Hong Kong residents. They occupy tiny spaces in multiresident housing, reduce waste by eating meals prepared efficiently in bulk, walk to most of their daily activities, use bicycles for primary transportation, and favor inherently low-carbon forms of recreation (sleeping late; engaging in pointless but environmentally nondestructive philosophical speculation; having sex). And recent college graduates are often quite green, too. Both my children, who are in their twenties, live without cars in small apartments in Manhattan. Neither chose urban living because of their father's arguments about energy and carbon; both were attracted, instead, by the density of excitement and opportunity which is a natural feature of any big, busy city. They spend a larger percentage of their income on housing than I do, but most of their other expenses are minimal—no car insurance, no swimming-pool maintenance, no driveway plowing—and if they want to see an interesting new movie they don't have to wait a month and drive twenty miles. If you're young and underpaid and living in a vibrant city, you have essentially unlimited opportunities for free or low-cost entertainment;* if you're young and underpaid and living in the suburbs, you spend a lot of time driving around or sitting at home in front of a screen.

Retired Americans often live low-impact lives as well. Seven years ago, my mother, who is now in her early eighties, moved into an apartment in a "continuing care retirement community." Her apartment is smaller than the semidetached house that she and my

* The classic work on this topic is *The Death and Life of Great American Cities*, by Jane Jacobs. See also Edward Glaeser's *Triumph of the City*, which was published in 2011.

father bought when they decided to downsize, several years after my brother graduated from college. She has her own kitchen but usually eats like a college student, in the facility's dining room or café. She still owns a car but puts many fewer miles on it than she did before she moved. And a number of her lifelong friends live down the hall. They regularly gather on foot for cocktails and meals, and they walk to their bridge games instead of driving. They live almost as efficiently as students, but with earlier bedtimes and quieter parties. And they don't face the looming crisis that haunts older people in the suburbs: the near total loss of freedom and independence which immediately follows the revocation of a driver's license.

In terms of reducing the environmental impacts of Americans, the confounding years are the ones between young adulthood and assisted living. The appeal of the suburbs can be impossible for young families to resist, assuming they can afford to flee, even though the main suburban family activity nowadays often seems to be driving bored children between widely separated activities. It's possible to arrange even nondense suburbs more compactly than we usually do—and, by doing so, to enhance rather than diminish the qualities that induce young families to spread out in the first place, while reducing some negative environmental and social effects—but, even in places where zoning rules don't stand in the way, history pulls strongly in the other direction. A big yard is very hard for an American with young children to resist, even if the children end up spending almost all their time indoors.

Young families, therefore, are unlikely to embrace urban density, as long as suburban sprawl remains affordable. Retirees, though, are another matter, since they constitute the demographic segment with the most to gain and least to lose from living compactly. If I were charged with slowing the American sprawl juggernaut, I might ignore the Sierra Club and cultivate an alliance with the AARP, which has 40 million members aged fifty and older—a potentially powerful force for environmental transformation. And, in fact, the AARP has become increasingly active in exactly that way.

11

Why Oil Is Worse Than Coal

It's usually said that coal is the worst of the fossil fuels, because burning it releases more carbon dioxide, sulfur dioxide, nitrogen oxide, carbon monoxide, mercury, arsenic, lead, cadmium, soot, and other pollutants, and because extracting and processing coal deposits can be horrifically destructive—as with "mountaintop-removal" coal mining, in which entire ecosystems are permanently ravaged. But oil is worse. The reason is that oil is the world's primary source of transportation fuels, and personal mobility underlies and exacerbates almost all the most tenacious environmental challenges, including the consumption of energy in other forms.

The critical environmental damage done by cars is

not caused by the fuel that they themselves consume, although they do plenty of that. (Direct fuel use by cars accounts for roughly a third of U.S. fossil-fuel use and carbon output.) The critical damage is caused by all the other consumption that driving fosters—consumption that would not occur on the same scale if drivers couldn't move around as easily as they do. Before cars, most people had to live close to other people and to the places where they worked and shopped, even if their homes were in small, isolated towns, far from other communities. Cars permanently changed that, by transforming the way their owners arrange themselves in relation to one another.

The major carbon-spewing energy drain in a sprawling American suburb isn't the car in the driveway; it's the driveway. That is, it's everything the car makes both possible and necessary: the oversized house, the three-bay garage, the manicured yard, the unused swimming pool, the miles of connecting asphalt, the redundant utilities, the schools, the hospitals, the shopping malls, and all the other accoutrements of inefficient suburban living—none of which would exist on anything like the same scale if residents were less able to move around at will. Cars are consumption amplifiers; driving is the pump that enlarges the sprawl balloon. And countries with rapidly modernizing economies, like China and India, are now following the American mobility example at extraordinary speed, by acquiring new cars and building new roads at a pace seldom matched even in the United States. It will be a while before those countries overtake Americans in impact per capita, but in absolute numbers they have already begun to make us look demure. And, as with us, the main driving-related environmental impacts will always be the indirect ones.

Yet most popular initiatives that supposedly address the environmental damage done by cars have the perverse effect of encouraging driving, by making car travel more pleasant, less expensive, or both: adding new travel lanes, managing roadways and intersections to make congested traffic flow more smoothly, using technology (including smartphone apps) to help drivers quickly find empty parking spaces, increasing miles per gallon, developing cars that run on cheaper fuels, giving drivers accurate real-time information about traffic bottlenecks (as with IBM's Smart Traveler system). Such efforts are ultimately counterproductive, from an environmental point of view, because they make drivers even happier with cars than they were already, and therefore exacerbate all the indirect impacts of driving, as well as making car owners less interested in real alternatives. The same is true of air travel, which also acts as a consumption amplifier by enlarging the scale of human interaction with the world's resources. Because I am a frequent air traveler, I own not only my house and the cars in my driveway but also what might be thought of as fractional shares in hotel rooms, rental cars, parking lots, restaurant tables, and innumerable other amenities, all over the world—my personal global timeshare infrastructure.

The two main forces with demonstrated effectiveness at truly reducing vehicle miles traveled—increased cost (as when oil prices spike or supplies are limited) and reduced convenience (as in rush-hour traffic jams or the creeping travel speeds on just about any side street in midtown Manhattan)—are so repellent to those of us who drive that when they emerge we usually demand relief. When U.S. gasoline prices rose to record levels in 2008, oil consumption fell by a significant amount—an environmental gain—but the main response, from voters and politicians, was to demand that oil be released from the Strategic Petroleum Reserve, that state fuel taxes be suspended, and that the drilling of oil wells be allowed in the Arctic National Wildlife Refuge. And we responded the same way when fuel prices rose again, in early 2011. We may say we're concerned about energy use and climate change, but when it comes to taking public action we Americans have shown repeatedly that what we mainly care about is how much we have to spend for gasoline—even though prices in the United States have never been more than about a half or a third of what they've been in Europe. Meanwhile, in our effort to get our stalled economy moving again, we have rescued struggling car manufacturers, offered subsidies to car buyers (who continue to favor oversized vehicles), and invested billions to repair and extend the infrastructure of driving.

12

Let Them Eat Kale?

In 2010, I gave a talk in Portland, Oregon, on the environmental value of urban density. Portland has a deserved reputation as a community with a high level of interest in environmental issues of all kinds, and it's the only large American municipality to have made a concerted, long-term, arguably effective effort to arrest the horizontal growth of its metropolitan area. That effort takes the form of Portland's "urban growth boundary," which divides the city proper from most of the rural and undeveloped land that abuts it. The boundary is intended to hold back the outward pressure of sprawling suburban subdivisions and their supporting infrastructure and to preserve agricultural land on the other side of the line.

Portland's boundary—which was mandated by a state requirement that applies to all Oregon municipalities—hasn't stayed in one place, and has always been controversial, and wouldn't be feasible if the metropolitan area hadn't been able, in effect, to sprawl across the Columbia River into Vancouver, Washington, which is next door. And Portland's residents, despite their widely admired light-rail system, are as dependent on automobiles as other Americans are, as you can easily observe for yourself while driving your rental car from the airport to your hotel downtown. Still, Portland remains the only truly significant American example of a large city attempting to accomplish by law what the densest American urban areas accomplished largely through accidents of geography and history. (Manhattan is compact mainly because it arose on a small island; downtown Boston and downtown San Francisco, which are probably the next two densest American urban areas, are on peninsulas.)

During my talk in Portland, the discussion turned to walkability, and to the characteristics a community needs to have if its residents are to enjoy the option of living with minimal dependence on cars. One member of the audience suggested that the most important factor might be distance from a decent grocery store. There are lots of other considerations—and for a good discussion of the most important ones, along with a very useful walkability calculator, see the Web site Walk Score (www.walkscore.com)—but distance to grocery stores is indeed critical. If you can walk to and from the place where you do most of your routine shopping, you can cut your driving significantly, and maybe even live without owning a car. And if there's a grocery store within walking distance of your home there are almost certainly other useful destinations nearby, too—clothing stores and restaurants and offices and public-transit stations. Connecticut's capital, Hartford, has been trying for years to revive its downtown. A recent initiative included the opening of a major grocery store, as a community anchor and magnet—a very good idea.

Once again, the key is density. When my wife and I lived in Manhattan, there were two full-sized grocery stores and a half dozen specialty food shops within a block of our apartment, and we were easily able to do all our food shopping on foot. That green convenience was made feasible by the local concentration of tall buildings filled with other grocery customers. The small town where I live now has a grocery store, but no one walks to it. And, because the store's customer base is small, its prices are high and the product selection is narrow, and those facts prompt many residents to do the bulk of their shopping in larger stores in neighboring towns, necessitating automobile round trips of many miles.

How far you live from your grocery store is of far greater environmental significance than how far you live from the places where your food is grown. This fact directly contradicts the main tenet of locavorism, which is the increasingly popular but misguided idea that it is environmentally irresponsible to eat food that was produced more than a short distance from your dining table. The number of miles that any food item travels from its origin to your plate is far less ecologically meaningful than how it was grown, what was sprayed on it while it was growing, how it traveled to market, and what else was traveling with it. A *New Yorker* colleague of mine raises chickens in her backyard, an increasingly popular hobby among people who want to be better environmental citizens. She enjoys the luxury of interesting breakfasts—and she gets to watch chickens through her kitchen window—but she also drives individual hens to the veterinarian, and that fact alone gives her eggs impressively huge carbon and energy footprints. A recent documentary about Portland's green consciousness shows a concerned resident driving her minivan twenty-five miles to buy two bags of fresh produce from a farmer on the other side of the city's urban growth boundary. And it shows the same farmer, in a pickup truck, transporting a larger but still very small selection of produce into the city, to sell at an urban farmers' market. Both trips are presented as virtuous acts, but neither makes environmental sense, as you can see easily if you think in terms not of distance but of energy and carbon expenditure per pound of delivered nutrients. If all the world's groceries traveled from farm to fork in minivans, two bags at a time, we'd have exhausted many of the world's resources long ago. Locavorism is appealing because, like many of the most popular green strategies, it feels enlightened yet entails no actual sacrifice, even if you don't grant yourself exemptions for coffee and out-of-season fruit.

The best book I've read about environmentally and morally responsible eating is *Just Food*, by James E. McWilliams, who is an associate professor of history at Texas State University at San Marcos and a former fellow of the Agrarian Studies Program at Yale. Among the many compelling points that McWilliams makes:

- Of all the energy expended in the production and processing of food, transportation represents the smallest share. Among the largest: home preparation, which accounts for a quarter of the total. As McWilliams observes, "we'd be better off focusing on what happens to our food after we buy it than on its place of origin." (Two especially energy-hungry food-preparation activities that are often celebrated as green: home canning and maple syrup making.)

- Grass-fed lamb imported to London from New Zealand has a quarter of the energy footprint of grain-fed lamb raised locally, in England; Brazilian apple juice shipped ten thousand miles to Germany represents a smaller expenditure of energy, overall, than the same product grown and processed locally. Such seeming anomalies are far from unusual. How food is grown is more telling than how far away.

- Organic farming, whatever its virtues, carries a surprisingly high environmental and human cost. Eschewing synthetic inputs results in lower yields and, therefore, encourages deforestation and "agricultural sprawl," which McWilliams defines as "an insidious form of development that threatens the world's remaining natural resources." Organic farming also almost always depends on large quantities of manure. Manure is less efficacious than synthetic fertilizers, and producing it in quantity requires large herds of defecating animals, which must themselves be fed, with the usual impacts. And many of the so-called natural chemicals commonly used by organic farmers—such as sodium nitrate, which is imported from South America—are more likely than synthetics to leach into watersheds and do other environmental harm.

- If you want to make a quick, large, permanent cut in your personal environmental impact, become a vegetarian. (According to my *New Yorker* colleague Michael Specter, "80 percent of all farmland is devoted to the production of meat," and farm animals are responsible for 20 percent of human-caused greenhouse-gas generation.) I myself haven't had much success at meatlessness yet, despite my tentative reclassification of bacon as a vegetable. But I'm sort of trying.

Food is hard to discuss rationally because, for most of us, meals are a highly emotional issue. My wife, Ann Hodgman, who has written several cookbooks, has observed that people tend to define themselves by what they refuse to eat: no meat, no broccoli, no pork, no genetically modified plants, no added sugars, "no hot foods that are mixed" (in the case of my nephews, when they were little). The environmental difficulty with locavorism is easier to see if you consider it in a less emotional context, by thinking about inedible agricultural products instead. Almost all of us wear at least some clothing made of cotton, and many people have argued that we should wear much more, since virtually all the synthetic alternatives are manufactured from fossil fuels. But would the environment be better off if we insisted on wearing only clothing made of cotton grown within fifty or a hundred or two hundred miles of where we live? Even in parts of the world that are well suited to cotton cultivation, successful crops require large quantities of pesticides, fertilizer, water, and fossil fuel. Growing cotton in places that are unsuited to it, solely for the purpose of shortening the distance between growers and consumers, would be an environmental disaster, not least because it would make extremely inefficient use of one of the world's most precious resources: high-quality arable land. The same difficulty applies to food crops. Transforming global agriculture in order to keep "food miles" at a minimum would inevitably mean increasing energy consumption, carbon output, water use, ecosystem devastation, and malnutrition. Locavorism, in that context, is mainly a luxury consumer choice made by people who have surplus disposable income. As a response to global hunger, it's the equivalent of "Let them eat cake."

A two-hundred-acre dairy farm near where I live has a sign that says "Every cow in this barn is a lady; please treat her as such." The farm's Web site, furthermore, says that its barns are designed "to cater to each cow's particular needs." The cows are never without human company and are regularly shampooed and vacuumed, and when one of them defecates a handler immediately cleans up after it. I've heard people describe milk from this farm as "sustainable," but it's anything but—as you can see if you scale the operation up to the size of the earth's human population. There are a number of other boutique dairy farms in my part of the country, and the milk they produce retails for as much as fifteen dollars a gallon. Food like that is a toy for rich people, not an environmental solution.

To say that the carbon and energy arguments for locavorism are flawed, or that organic agriculture is not sustainable for an earth-sized population, is not to endorse factory farming or the grotesque diet of the average American consumer. (This is a complex issue that McWilliams, who is a vegan, is especially good at untangling, although his answers will not delight agricultural romantics.) And enchantment with locally grown food can have environmental value—though not in the way people tend to think. There's a farmers' market in my daughter's neighborhood in Manhattan, down the street from her apartment. Like most such markets, it has bigger carbon and energy footprints than Gristedes, the Food Emporium, or any other of the city's regular grocery stores, for all the reasons McWilliams describes. But having such a market nearby makes my daughter and many of her neighbors happier about living where they do—and that's good for the environment because the contentment of urban dwellers has environmental value. Still, growing crops on the roofs of apartment buildings, in downtown vacant lots, or in skyscraper-sized urban "vertical farms"—three popular, supposedly eco-friendly ideas—makes no sense from an environmental point of view. There's a powerful back-to-the-land strain in American environmentalism, and all such schemes draw their power from it. But the underlying idea is fallacious. There's nothing green about trying to make cities more like the country. On the contrary.

13

Traffic Congestion Is Not an Environmental Problem

The photograph on the following page was taken in Tampa, Florida, in the spring of 2001, as part of a campaign by Hillsborough Area Regional Transit to build support for a proposed light-rail project. (That effort was abandoned in 2005; HART is now trying again.) The shot was staged, but it's a fair depiction of a typical American rush-hour traffic jam: four lanes of bumper-to-bumper cars, frozen in place, spewing exhaust. Such congestion is increasingly common. In 1949, just 3 percent of American households owned more than one car; today, we have 50 million more registered automobiles than licensed drivers, and the average passenger vehicle (according to the Environmental Protection Agency)

Hillsborough Area Regional Transit

travels twelve thousand miles a year. And the rest of the world is catching up. China, which had very few private automobiles just twenty-five years ago, is now the world's largest market for new cars.

The photo on page 82 shows all the occupants of the cars in the first photo, sitting on folding chairs in the middle of the same street. Even to someone who has spent time thinking about traffic and transit, the second photo is a shocker: is that really *all* the people those cars were carrying? The first photo looks like an entire suburb's worth of commuters; the second looks like about half the occupants of a single city bus.

These two pictures powerfully make the point that

Hillsborough Area Regional Transit

HART officials hoped they would: automobiles are an absurdly wasteful form of transportation. (A typical SUV driver constitutes just 2 or 3 percent of the total load being moved by his SUV; the rest is SUV.) When I drive on a highway now, I sometimes mentally superimpose these images on the traffic ahead and re-astonish myself at the profligacy of solo mobility. It takes a big crowd on a wide road to add up to more passengers than the maximum occupancy of a single bus or train car.

But the deeper environmental significance of the Tampa photographs is not the simplistic notion that cars are bad and transit is good. The reason is that, if you're a car driver who spends time in traffic jams like the one in the first photo, the second photo looks like Nirvana to you, because it shows a formerly exasperating stretch of road that now contains plenty of room for you and your car: problem solved! Drivers hate congestion because it makes driving slow, aggravating, and boring; moving competing cars off the road makes drivers happy about driving again—and that's not an environmental gain.

Building good transit isn't a bad idea, but it can actually backfire if the new trains and buses merely clear space on highway lanes for those who would prefer to drive—a group that, historically, has included almost everyone with access to a car. To have environmental value, new transit has to replace and eliminate driving on a scale sufficient to cut energy consumption overall. That means that a new transit system has to be backed up by something that impels complementary reductions in car use—say, the physical elimination of traffic lanes or the conversion of existing roadways into bike or bus lanes, ideally in combination with higher fuel taxes, parking fees, and tolls. Needless to say, *those* ideas are not popular. But they're necessary, because you can't make people drive less, in the long run, by taking steps that make driving more pleasant, economical, and productive.

* * *

One of the few forces with a proven ability to slow the growth of suburban sprawl (other than mortgage-market implosions) has been the ultimately finite tolerance of commuters for long, annoying commutes. That tolerance has grown in recent decades, and not just in the United States, but it isn't unlimited, and even people who don't seem to mind spending half their day in a car eventually reach a point where, finally, enough is enough. That means that traffic congestion can have environmental value, since it lengthens commuting times and, by doing so, discourages the proliferation of still more energy-hungry subdivisions—unless we make the congestion go away. If, in a misguided effort to do something of environmental value, municipalities take steps that make long-distance car commuting faster or more convenient—by adding lanes, building bypasses, employing traffic-control measures that make it possible for existing roads to accommodate more cars with fewer delays, replacing tollbooths with radio-based systems that don't require drivers even to slow down—we actually make the sprawl problem worse, by indirectly encouraging people to live still farther from their jobs, stores, schools, and doctors' offices, and by forcing municipalities to further extend road networks, power grids, water lines, and other civic infrastructure. If you cut commuting time by 10 percent, people who now drive fifty miles each way to work can justify moving five miles farther out, because their travel time won't change. This is how metropolitan areas metastasize. It's the history of suburban expansion.

Traffic congestion isn't an environmental problem; traffic is. Relieving congestion without doing anything to reduce the total volume of cars can only make the real problem worse. Highway engineers have known for a long time that building new car lanes reduces congestion only temporarily, because the new lanes foster additional driving—a phenomenon called induced traffic. Widening roads makes traffic move faster in the short term, but the improved conditions eventually attract additional drivers and entice current drivers to drive more, and congestion reappears, but with more cars—and that gets people thinking about widening roads again. Moving drivers out of cars and into other forms of transportation can have the same effect, if existing traffic lanes are kept in service: road space begets road use.

The ideal automobile strategy would be to steadily remove driving lanes while maintaining congestion at levels that drivers find vexing, thereby giving them an ongoing incentive to embrace alternatives. If Tampa ultimately succeeds in funding its proposed light-rail system—probably still a long shot—will city leaders find the courage to complete the project by eliminating enough existing traffic lanes to force current drivers to travel by rail instead? Probably not. (In 2011, a relatively modest effort by New York City to turn car lanes into bicycle lanes resulted in a lawsuit by a group of Brooklyn residents.) One of the arguments that cities inevitably make in promoting transit plans is that the new system, by relieving automobile congestion, will improve the lives of those who continue to drive. No one ever promotes a transit scheme by arguing that it would make traveling *less* convenient—even though, from an environmental perspective, inconvenient travel is a worthy goal.

14

Transit That's Bad for the Environment

In the summer of 2010, what is said to have been the largest traffic jam in the history of the world arose in northern China, primarily along portions of Highway 110, which runs for more than eight hundred miles between Beijing and Yinchuan. The main components of the slowdown were trucks that were transporting coal from mines in Inner Mongolia, but the country's burgeoning national fleet of passenger vehicles contributed. The main blockage stretched more than sixty miles and lasted almost two weeks. Vehicles advanced as little as a kilometer a day. Truck drivers slept on the roadway, under their trailers.

One cause of the stoppage was construction of the highway itself. In 1988, less than twenty-five years ago,

China had no automobile expressways. Since then, it has laid thousands of miles of asphalt, but the construction still hasn't kept up with the explosive increase in the number of Chinese drivers. Peter Hessler—a *New Yorker* colleague of mine and the author of several terrific books about China, the most recent of which is *Country Driving*—attended a press conference in Beijing in 2005 at which the Ministry of Communications announced that, by 2020, China intended to have more highway miles than the United States. Hessler told me, "The plan is to connect every city with a population of more than 200,000."

China's transformation from a nation of walkers and bicycle riders into a nation of eager, mostly clueless automobile drivers has been breathtaking. I visited Beijing in 2006 and spent a lot of time walking around the city. One evening—after stepping in wet concrete while attempting to take a shortcut through a huge construction site, and then realizing I didn't know how to cross a major new expressway on foot—I decided to take a taxi the rest of the way back to my hotel. The cab I found was in a line in front of the Place, a brand-new, 700,000-square-foot shopping mall and office development. The Place's signature feature is an enormous entrance canopy whose ceiling is one of the world's largest television screens: a five-section LED display more than eight hundred feet long and almost a hundred feet wide. (The display cost $32 million and was designed by the American company that also designed the world's largest television screen, which is at the Fremont Street Experience in Las Vegas.) Once I was in the cab, I leaned back in the seat, closed my eyes, and dozed: it was rush hour, and the traffic was murder. After about half an hour, I opened my eyes again and looked around to see if I could tell where we were. And directly above us (though now on the left side of the cab, instead of the right) was the monumental television screen at the Place. It had taken the driver half an hour to inch down to where he could make a U-turn and then inch back up to roughly where we had started.

One way the Chinese have dealt with snarled traffic like this has been by bulldozing ancient neighborhoods to make room for more expressways. Even so, the country is producing drivers faster than pavement. A Chinese company now offers a traffic-jam-rescue service: you call from your immobilized car, and a guy on a motorcycle brings a substitute driver to take your place, then carries you to your destination on the back of his motorcycle. Your car catches up with you later.

Another novel "solution" to the Chinese car problem is

a concept transit vehicle called the 3D Fast Bus, also known as the straddling bus (see an artist's conception on the following page). The bus would carry as many as 1,200 passengers, span two traffic lanes, and be tall enough so that cars and small trucks could drive underneath it yet not so tall that it couldn't fit under bridges and overpasses. It would be served by elevated boarding stations and would be able to load and unload passengers without impeding the flow of car traffic, which would pass underneath, as through a tunnel. The bus's developer—the Shenzhen Huashi Future Parking Equipment Company—presented scale models and a computer simulation during the summer of 2010, shortly before the emergence of the big Highway 110 traffic jam, and the company is supposedly looking for partners to help it bring the concept to other countries, among them the United States.

The straddling bus has been described, by its inventor and others, as an environmental breakthrough. The company has said that each bus would be significantly more fuel efficient than other forms of transportation and would reduce the world's carbon output by more than twenty-five hundred tons a year, in comparison with whatever the alternatives might be. The vehicle wouldn't require tunnels (unlike a subway system), elevated tracks

Shenzhen Huashi Future Parking Equipment Company

(unlike a monorail), or traffic lanes from which cars would be excluded (unlike bus rapid transit). It would require only a relatively small investment in additional infrastructure, including tracks or paths for its wheels, as well as a system for supplying electric current. (Some of the conceptual drawings show rooftop photovoltaic panels, but, given the size of the vehicles and their intended load and function, those panels must be mainly for show.) The general reaction on green-oriented U.S. Web sites was that here, finally, was a cool-looking solution to traffic jams; a mention in the *Huffington Post* called the straddling bus potentially "an environmentally friendly way to save money while easing congestion on city roads." *Time* picked it as one of "The 50 Best Inventions of 2010."

But the idea's main appeal—that it doesn't get in the

way of cars—is also its main flaw. And in that way the straddling bus is like most proposed "solutions" to congestion: it accommodates drivers—like the emptied commuter street in the second Tampa photograph—rather than forcing anyone to drive less. A far greener transportation solution would be a system of transit vehicles that actually displaced drivers—perhaps just regular buses operating at ground level in repurposed car lanes on existing roadways. But something like that wouldn't seem nearly as interesting, futuristic, or convenient. And it would make drivers furious.

My state, Connecticut, is currently planning a dedicated commuter-bus line, called the Busway, on a former railroad bed that runs between Hartford and New Britain, a distance of a bit less than ten miles. The project is projected to cost $569 million (or almost $60 million a mile), of which almost half is expected to be paid by the federal government. As might be expected, Connecticut's governor has pushed the Busway not only as a "green" initiative but also as a boon to drivers, since, like the straddling bus, it will alleviate (temporarily) the traffic jams that currently annoy commuters on I-84, which connects the same cities—the same sort of magical thinking that infects almost all such projects.

15

Fast Trains and the Prius Fallacy

A couple of years ago, I took part in a land-use seminar in Cambridge, Massachusetts. One of the speakers was Michael Dukakis, the state's former governor and the 1988 Democratic candidate for president. Dukakis's topic, about which he has been evangelical, was high-speed rail—trains that travel fast enough to compete not just with cars but also, often, with airplanes. Like other proponents, including President Obama, Dukakis (who has described himself as a rail fanatic) spoke of fast trains as a commonsense green solution to some of the country's energy and emissions problems, as well as a job creator and promoter of economic health.

I've traveled on a few high-speed trains: the Eurostar

between London and Paris; the Alfa Pendular between Coimbra and Porto, in Portugal; and the Amtrak Acela between New Haven, Connecticut, and Washington, D.C. All three trains were impressively fast—especially the Alfa Pendular, which reached speeds of over two hundred kilometers an hour, according to the digital display in my car. The Acela is slower, but the ride is quite pleasant, especially in comparison with the hellish, survival-oriented form of driving that's usually required on the New Jersey Turnpike. And the Eurostar, though not as speedy as the Portuguese train, makes the trip between London and Paris in under two and a half hours, with far less hassle than flying.

The agreeableness of riding on high-speed trains is a big part of their appeal. But it's also an environmental problem, because one consequence of making long-distance travel cheaper and more convenient is that it causes people to want to do more of it. (This is an issue with all mobility enhancers, including cheap gasoline, discount airlines, and frequent-flier miles.) The Eurostar made it easy for me, when I was in London, to hop over to Paris for lunch and the latest exhibit at the Louvre, ho-hum. The environment would have been better off if, instead of doing that, I had stayed in London and walked from my hotel to the British Museum. And it would have been better off still if I'd skipped London altogether and stayed at home, on my side of the Atlantic.

In the United States, the main environmental benefit of travel by train, versus travel by air, is not that it consumes less energy per passenger per mile (a point that's debatable, anyway, since the supposed energy and carbon advantages of train travel depend on several speculative factors, including average passenger loads that are much higher than the ones that rail romantics envision when they daydream about the pleasures of train travel); the main environmental benefit of train travel is that it takes longer, making people less likely to do it. If the only way to cross the country were to take a train, no New Yorker would ever travel to Las Vegas for the weekend. That would be good for the environment, though bad for Las Vegas (yet good for Atlantic City?). Rail enthusiasts are highly susceptible to what might be called the Prius Fallacy: a belief that switching to an ostensibly more efficient travel mode turns mobility itself into an environmental positive. Europeans use less energy in moving around than Americans do, but the reason isn't that they do more of their traveling by train; the reason is that their usual destinations are more tightly clustered than Americans' usual destinations are—because Europe covers less area and is more densely populated—so their average trips are shorter, no matter which mode they use. And the environment would be better off if they stayed home, too, their speedy trains notwithstanding.

My state is considering high-speed rail: a new line along a sixty-two-mile existing rail route between New Haven, Connecticut, and Springfield, Massachusetts. The new line, which won a $40 million federal stimulus grant in 2010 and is expected to cost something like $900 million, will shorten travel times between those cities. The putative environmental benefits are all the standard ones: less gasoline, less exhaust, less carbon, less congestion. The actual environmental downside is that, if the line ever does get built, it will enhance the appeal of long-distance commuting and (therefore) act as an accelerant to sprawl, especially for people who don't mind living in one state and working in another. It will also—if it attracts a significant number of travelers from Interstate 91, which connects the same cities but suffers congestion during rush hours—make driving the same route faster and more convenient than it is currently, for those who don't want to take the train. In that way, the train line, if it's actually used, would eventually exacerbate the environmental problems it was built to solve.

In 2011, Florida's governor, Rick Scott, turned down $2.4 billion in federal funding for a proposed high-speed rail line between Orlando and Tampa, a distance of eighty-four miles. Scott cited the likely long-term cost to Florida, which, even more than most states, was already struggling with budget difficulties. And, he pointed out, the train, even if it functioned as promised, would shorten the trip between the two cities by only half an hour in comparison with traveling by car. (And that's assuming the train's schedule aligned perfectly with the desires of travelers—a big stretch.) Scott, if he had chosen to, could have made a green argument as well: every car driver that the proposed train succeeded in luring away from Interstate 4 would reduce congestion on the same highway, thereby making driving more appealing than it is now and, very possibly, more appealing than taking the train. Besides, what would all those train passengers be likely to do once they arrived in Orlando or Tampa? Walk?

Rail proposals, for those who love them, are attractive for several reasons, in addition to the supposed environmental ones. One is that, like all big infrastructure efforts, they call for *more*, and shopping for more is always exciting. Another is that high-speed trains are technologically advanced, and we know that technology is the answer to our problems. (Especially enticing are proposals for systems that would require huge research and construction investments—such as maglev trains, which are levitated above their tracks by magnets, and "vactrains," which are maglev trains that operate in vacuum tubes, to eliminate wind resistance, and are theoretically capable of traveling as fast as commercial jets.) Another attraction is that building and operating railroads puts people to work—an economically and socially valuable thing, especially during times of high unemployment. Another is that trains—even futuristic ones—have a powerful nostalgic appeal.

But all these are really fantasies. Building railroads does create jobs, but if the railroads themselves don't make long-term sense it would be smarter to put people to work doing something with actual environmental value. And the romantic appeal of trains is far stronger in the abstract than it could ever be in reality. In fact, for most travelers, the rapture wouldn't survive a single round trip between Boston and San Francisco. As with cars, planes, and public transit, we don't need *more* transportation choices. We need fewer choices—but better ones. And if we're serious about confronting our energy and climate problems we need less transportation overall.

16

Increased Efficiency Is Not the Answer

In April 2010, the federal government adopted standards for automobiles requiring manufacturers to improve the average fuel economy of their new-car fleets by 30 percent, to 35.5 miles per gallon, by 2016. The *New York Times*, in an editorial titled "Everybody Wins," said the change would produce "a trifecta of benefits." Those benefits had been enumerated the year before by Steven Chu, the secretary of energy: a reduction in total oil consumption of 1.8 billion barrels; the elimination of 950 million metric tons of greenhouse-gas emissions; and savings, for the average American driver, of three thousand dollars over the life of each new car.

Chu, who shared the Nobel Prize for Physics in 1997,

has been an evangelist for energy efficiency, and not just for vehicles. He has often said that, as far as efficiency is concerned, there isn't simply low-hanging fruit still to be picked; there's fruit actually lying on the ground. I spoke with him in July 2010, shortly after he'd conducted an international conference, in Washington, called the Clean Energy Ministerial, at which efficiency was among the main topics. "I feel very passionate about this," he told me. "We in the Department of Energy are trying to get the information out that efficiency really does save money and doesn't necessarily mean that you're going to have to make deep sacrifices. It just means being smarter about things."

Efficiency has been called "an invisible powerhouse" and "the fifth fuel"; it is seen as a cost-free tool for accelerating a transition to a green-energy economy. In 2007, the United Nations Foundation said that efficiency improvements constituted the world's "largest, most evenly geographically distributed, and least expensive energy resource." In 2009, the management-consulting firm McKinsey concluded that a national efficiency program could "reduce end-use energy consumption in 2020 by 9.1 quadrillion BTUs, roughly 23 percent of projected demand, potentially abating up to 1.1 gigatons of greenhouse gases annually." Amory Lovins, whose thinking has influenced Chu's, has written and lectured about efficiency for many years and has served as an energy consultant to many corporations. He has referred to the replacement of incandescent lightbulbs with compact fluorescents as "not a free lunch, but a lunch you're paid to eat," since a fluorescent lamp will usually save enough electricity to more than offset its higher purchase price. Lovins also coined the term "negawatt" to describe energy reductions achievable by making powered devices more efficient or by disconnecting them when they're unneeded. Tantalizingly, much of the technology required to harvest negawatts is well understood. In 2010, the World Economic Forum, in a report called "Towards a More Energy Efficient World," observed that "the average refrigerator sold in the United States today uses three-quarters less energy than the 1975 average, even though it is 20% larger and costs 60% less in inflation-adjusted terms"—an improvement that Chu also cited in his conversation with me.

But the issue is less straightforward than it may seem. Historically, economy-wide energy savings from efficiency gains have been hard to pin down, to say the least. That thirty-five-year period during which new refrigerators have plunged in electricity use and price is also a period during which the global market for refrigeration has burgeoned and the world's total energy consumption and carbon output, including the parts directly attributable to keeping things cold, have climbed. Similarly, the first fuel-economy regulations for U.S. cars—which were implemented in 1975 in response to the Arab oil embargo of 1973 and 1974—were followed not by a steady decline in total U.S. motor-fuel consumption but by a long-term rise, as well as by increases in horsepower, curb weight, vehicle miles traveled (up 100 percent since 1980), and car ownership.

Most economists and efficiency experts would argue that such increases in consumption are merely a result of rising incomes and economic growth, and that the correlation between increased efficiency and mounting energy use implies nothing about causation. But a growing group of economists and others have argued that the association is more than coincidental. They have said, in fact, that improvements in energy efficiency can exacerbate the problems they are meant to solve, more than negating any environmental gains—an idea that was first proposed a century and a half ago and that came to be known as the Jevons paradox.

17

William Stanley Jevons

Great Britain in the middle of the nineteenth century was the world's leading military, industrial, and mercantile power. Its fleets were unmatched, its colonial empire was the world's largest, and its factories were rapidly extending the Industrial Revolution, which they had inaugurated. In 1865, a twenty-nine-year-old Englishman named William Stanley Jevons published a book, *The Coal Question*, in which he argued that the bonanza couldn't last. Britain's affluence and global hegemony, he wrote, depended on its endowment of coal, which the country was rapidly depleting. He added that such an outcome could not be delayed through increased "economy" in the use of coal—what we refer to today as energy efficiency. He concluded, in italics, *"It is wholly a confusion of ideas to suppose that the economical use of fuel is equivalent to a diminished consumption. The very contrary is the truth."*

He offered the example of the British iron industry. If some technological advance made it possible for a blast furnace to produce iron with less coal, he wrote, then profits would rise, new investment in iron production would be attracted, and the price of iron would fall, thereby stimulating additional demand. Eventually, he concluded, "the greater number of furnaces will more than make up for the diminished consumption of each." The British railway system, which was still young, provided a more complex example of how efficiency and consumption can grow in tandem and reinforce each other. As Jevons pointed out, Britain's railways evolved outward from its coalfields, with the effect that the entire system "converges upon them, just as the arteries and veins of the animal body converge upon the heart and lungs." James Watt developed his coal-powered steam engine primarily to pump water from British coal mines, and a more efficient version of the same engine powered the first locomotive, which was invented specifically to carry coal more rapidly and inexpensively from a British colliery to a coastal shipping port. That locomotive ran on rails that were made of iron, which had been forged in furnaces powered by coal. And, as the use of trains grew, a mounting need for rails increased the demand for coal, and that stimulated a boom in mining and a consequent drop in price—and the resulting coal bonanza pushed down the cost of rails, as well as increasing demand for train cars and (iron) railway bridges. The rise of coal also led directly to the commercial production of coal gas and to the manufacturing of (iron) piping to carry coal gas into homes. This swirling, self-reinforcing multiplier effect continued long after Jevons's death and reached corners of the global economy that, at first glance, seem wholly unrelated to it. Several of the most famous golf courses in the world were built in Scotland and England in the late nineteenth century by British railway companies, whose goal was to induce Britons to venture farther from home by train. That led to a British golf boom—the game rapidly became so popular that players at Musselburgh, Scotland, sometimes teed off a dozen or more at a time—and the boom spread across the Atlantic, as emigrating Scottish professionals carried golf to the United States. Many of the oldest American golf clubs arose as a result: St. Andrews, in Yonkers, New York; Shinnecock Hills, on

Long Island; Chicago Golf Club; Van Cortlandt Park, in the Bronx (the first public course in the United States, founded in 1895); the little nine-hole club in my town in Connecticut. All of those golf clubs can be considered indirect international by-products of an efficiency-fueled late-nineteenth-century surge in Britain's production of coal, iron, and trains. And the reverberations continue today, as (for example) real-estate developers in China, Vietnam, Morocco, Russia, and numerous other unlikely locales build golf resorts in the hope of attracting golf-obsessed business travelers from the United States, Australia, the United Kingdom, Korea, and Japan.

Other examples of the association between increased efficiency and increased consumption abound. In a paper published in 1998, the Yale economist William D. Nordhaus estimated the cost of lighting throughout human history. An ancient Babylonian, he calculated, needed to work more than forty-one hours to acquire enough lamp oil to provide a thousand lumen-hours of light—the equivalent of a standard seventy-five-watt incandescent lamp burning for a little less than an hour. Thirty-five hundred years later, a contemporary of Thomas Jefferson's could buy the same amount of illumination, in the form of tallow candles, by working for about five hours and twenty minutes. By 1992, an average American, with access to compact fluorescents, could do the same in less than half a second.

In other words, increasing the energy efficiency of illumination is nothing new; improved lighting has been "a lunch you're paid to eat" ever since humans upgraded from cave fires (fifty-eight hours of labor for our early Stone Age ancestors, according to Nordhaus). Furthermore, the effect is even larger than it seems, because our ever-growing ability to inexpensively illuminate our activities has transformed our lives in ways that go far beyond our expenditures on lighting. Increasingly inexpensive, efficient illumination has lengthened the workday, increased our opportunities for energy-hungry leisure, and given us access to luxuries that would otherwise be inconceivable. Many sources of artificial light—our television sets, our computer screens, our mobile telephones, the control panels of our appliances, the front panels of vending machines, the projectors in movie theaters, the signs and billboards along our highways, the slot machines in Las Vegas casinos—don't even register in our minds as forms of illumination. Indeed, we now generate light so extravagantly, and at so little incremental cost, that darkness itself has become an endangered natural resource.

Again, correlation proves nothing about causation. Nor is there anything earthshaking in pointing out that people nowadays are wealthier and consume more than people in the past. Yet it's also apparent that sustained, dramatic improvements in the efficiency of lighting have not caused a drop in the total amount of energy we expend on illumination, or shrunk energy consumption overall—a fact that, at the very least, ought to make us skeptical about predictions that further efficiency gains will cause global energy consumption to fall.

The pressing question is whether the synchrony between efficiency and consumption is a trivial one or whether, as Jevons believed, it reflects a fundamental economic and behavioral truth—and that is an issue which clearly bears on personal, national, and global policy choices regarding energy use and climate change and is therefore even more important today than it was when Jevons raised it.

18

The Coal Question

Jevons was born in Liverpool in 1835. His mother died when he was ten, and his father, an iron merchant, was financially ruined three years later, in the collapse of the British railway bubble. Jevons spent two years at University College, in London, and then, at the age of nineteen, emigrated to Australia, where he had been offered a job as an assayer at a new mint in Sydney—an opportunity created by Australia's gold rush, which was roughly concurrent with ours. He became an avid and skilled photographer, and he wrote an authoritative book on the climates of Australia and New Zealand. He left Australia after five years, visited a brother in the United States, completed his education in England, published a well-received book on the price of gold, and became a part-time lecturer in logic and political economy. *The Coal Question* made him a minor celebrity; it was praised by John Stuart Mill and William Gladstone, and it inspired the government to commission a study. In 1869, Jevons—who loved both calculus and the work of the British mathematician George Boole—invented a mechanical "logic machine," which looked a little like an upright piano and was a distant ancestor of the computer. Two years later, he published *The Theory of Political Economy*, a book that's still described as one of the founding texts of mathematical economics. He drowned in 1882, at the age of forty-six, while swimming in the English Channel. In 1905, John Maynard Keynes, who was then twenty-two and a graduate student at Cambridge University, wrote to Lytton Strachey that he had discovered a "thrilling" book: Jevons's *Investigations in Currency and Finance*. Keynes wrote of Jevons, "I am convinced that he was one of *the* minds of the century." Later, Keynes described Jevons as "the first theoretical economist to survey his material with the prying eyes and fertile, controlled imagination of the natural scientist."

William Stanley Jevons

Mill and Keynes notwithstanding, Jevons might be little discussed today, except by historians of economics, if it weren't for the scholarship of another English economist, Len Brookes. At the time of the 1970s oil crisis, Brookes was working for the United Kingdom Atomic Energy Authority. He argued that devising ways to produce goods with less oil—an obvious response to higher prices—would merely accommodate the new prices, causing energy consumption to be higher than it would have been if no effort to increase efficiency had been made; only later did he discover that Jevons had anticipated him by more than a century. I spoke with Brookes in 2010. He told me, "Jevons is very simple. When we talk about increasing energy efficiency, what we're really talking about is increasing the productivity of energy. And, if you increase the productivity of anything, you have the effect of reducing its implicit price, because you get more return for the same money—which means the demand goes up."

Nowadays, Jevons-type effects are usually referred to as "rebound"—or, in cases where increased consumption more than cancels out any energy savings, as "backfire." In a 1992 paper, Harry D. Saunders, an American researcher, provided a concise statement of the basic idea: "With fixed real energy price, energy efficiency gains will increase energy consumption above where it would be without these gains." Saunders has explored the subject extensively since then. In 2011, he wrote, "Rebound effects are subtle and complex. And they create some thorny issues for policy makers that will need to be addressed. Most importantly, energy analysts must cease utilizing a far-too-simple assumption that efficiency gains yield direct and linear reductions in energy use. The complex and varied economic phenomena known collectively as 'rebound effects' mean that we cannot expect that improving the energy efficiency of steel production by 30 percent, for example, will yield a simple and direct 30 percent reduction in the energy consumed by the steel sector, let alone the economy as a whole. Just as economists expect that gains in labor productivity will contribute to greater employment overall, not less, gains in energy productivity (a.k.a. energy efficiency) are not likely to be taken up simply as direct reductions in energy demand overall."

In 2000, the journal *Energy Policy* devoted an entire issue to rebound. It was edited by Lee Schipper, who, at the time, was working at the International Energy Agency, in Paris, on leave from the Lawrence Berkeley National Laboratory, which is run by the U.S. Department of Energy. (Schipper, who was a generous scholar and gifted teacher, died of pancreatic cancer, in 2011, at the age of sixty-four.) In an introductory editorial, Schipper wrote that the question was not whether rebound exists, but, rather, "how much the effect appears, how rapidly, in what sectors, and in what manifestations." The *Energy Policy* contributors included Brookes, Saunders, and an international selection of experts, a majority of whom concluded that there wasn't a lot to worry about. Schipper, in his editorial, wrote that the articles, taken together, suggested that "rebounds are significant but do not threaten to rob society of most of the benefits of energy efficiency improvements."

I spoke with Schipper in 2010, when he was a senior research engineer at Stanford University's Precourt Energy Efficiency Center, and he told me that the Jevons paradox has limited applicability today. "The key to understanding Jevons," he said, "is that processes, products, and activities where energy is a very high part of the cost—in this country, a few metals, a few chemicals, air travel—are the only ones whose variable cost is very sensitive to energy. That's it." Jevons wasn't wrong about nineteenth-century British iron smelting, he said; but the young and rapidly growing industrial world that Jevons lived in no longer exists. Later, Schipper added that associations like the one between increased lighting efficiency and increased energy consumption not only prove nothing about causation, they also "say nothing about how much energy would have been consumed had efficiency *not* increased." He continued, "The correct comparison is not between before and after; rather, it's between with efficiency improvements and without, with due accounting for possible changes in energy prices."

Counterfactuals like this one are impossible to know, of course—would Australia still be a popular international travel destination if no one had invented the passenger jet?—but most economists and efficiency experts, after studying modern energy use, have come to similar conclusions. It's common to say, for example, that when you increase the fuel efficiency of cars you lose no more than about 10 percent of the fuel savings to increased use: a car's owner will respond to the drop in per-mile cost by driving a bit more—but only a little bit more. And studies of other energy-using devices have described effects that are smaller still. Amory Lovins, in a response to a 2010 *New Yorker* article of mine on this subject, wrote, "Rebound effects are small for three reasons: no matter how efficient your house or washing machine becomes, you won't heat your house to sauna temperatures, or rewash clean clothes; you can't find an efficient appliance's savings in your un-itemized electric bill; and most devices have modest energy costs, so even big savings look unimportant." And Schipper told me that, if you look at the whole economy, rebound effects are comparably trivial. He said, "People like Brookes would say—they don't quite know how to say it, but they seem to want to say the extra growth is more than the saved energy, so it's like a backfire. The problem is, that's never been observed at the national level." Lovins, furthermore, has said, "The rebound argument is rediscovered about every decade; this is about the fourth round I can recall. It grows no more convincing with age. Essentially it's a *post hoc ergo propter hoc* logical fallacy [i.e., mistaking correlation for causation], contending that since we've been trying to save energy, yet its use keeps growing, therefore its growth must be due to our trying to save it. I can imagine few false contentions whose adoption would be less likely to advance the human prospect."

But troublesome questions have lingered. In 2004, a committee of the House of Lords invited a number of experts to help it grapple with a puzzling issue: the United Kingdom, like a number of other countries, had invested heavily to increase energy efficiency in an attempt to reduce its greenhouse emissions. Yet energy consumption and carbon output in Britain—as in the rest of the world—had continued to rise. Why?

19

How Increasing Efficiency Causes Overall Energy Consumption to Rise

Harry Saunders, in a response to Amory Lovins's objections to my *New Yorker* article, told me, "In Amory's world, energy efficiency gains reduce energy use in a one-for-one fashion and everything else stays essentially the same as it would otherwise be. His 'hoc' (energy efficiency) accordingly leads him to a 'propter hoc' that is wildly off the mark. In reality, efficiency gains in the productive part of the energy economy unleash a host of effects that travel via multiple pathways through the economy: substitutions; output price reductions that feed other producers and work their way through products and services to the final consumer; newly enabled applications and products; and so forth. In economist-speak, Amory makes the mistake of treating energy-cost minimization as the economic force, when in fact it is overall profit-maximization and consumer welfare–maximization, which lead to far different energy-use trajectories. Actual energy-use dynamics are much more complex than his characterization, however alluring his argument may appear to the rational mind. Unfortunately, he's simply wrong."

Most economic analyses of rebound, like the ones that Lovins cites, focus on particular uses or categories of uses: if people buy a more efficient clothes dryer, say, what will happen to the energy they use as they dry clothes? (At least one such study has concluded that, for appliances in general, rebound is nonexistent.) But Brookes and Saunders, among others, have argued that studies like that one—which Brookes refers to as "bottom-up"—are inherently flawed, because they sample only a very small part of the relevant economic picture and ignore or understate the biggest consumption effects, which occur in economies as a whole. In fact, Brookes dislikes the term "rebound," which he feels trivializes the real impacts by making efficiency analysis seem like a simple one-input/one-output problem. The energy effects that he and Saunders and others are talking about are actually less like a bounce than like a chain reaction—perhaps a better term.

Saunders, in 2011, wrote, "Many of us, when contemplating the potential for reduced energy use, quite naturally reference our thoughts around those opportunities we see around us in our personal realm—energy used in the household or for private transportation. But this 'end use' energy consumption represents only a relatively small fraction of the energy we actually consume. Globally, some *two-thirds* of all energy that is consumed is the energy used to produce the goods and services we consume. Your washing machine may be very efficient in its use of energy, but think of the metal body alone and the energy required to mine, smelt, stamp, coat, assemble, and transport it to the dealer showroom where you bought the appliance. The energy 'embedded' in your washing machine is substantial. The same is true for any product you purchase or service you consume."

Even with end-use energy consumption, the long-term effects aren't always tiny. When OPEC raised oil prices in the 1970s, it did the economic equivalent of imposing a carbon tax on petroleum. One result was that U.S. oil consumption fell significantly, as Americans, in an effort to save money, cut back on driving and took other energy-saving steps. And many Americans, over the next few years, switched to more fuel-efficient cars, both on their own initiative and as a result of more stringent fuel-economy standards imposed by the government. But those efficiency gains, rather than causing a permanent decline in energy use, had the effect that Len Brookes said they would: they accommodated the higher oil price—in effect, rescinding the OPEC carbon tax—and alleviated the economic discomfort that had given Americans an incentive to think about reducing their profound dependence on cars. And, as the actual price of oil stabilized and even fell, fuel consumption resumed its upward rise with a new impetus, both per capita and overall. Automobiles actually continued to become more efficient in dozens of ways; but, in the absence of continuing pressure from the price of oil, those efficiency gains mainly took the form not of additional miles per gallon but of energy-hungry features like greater vehicle weight, increased horsepower, more air-conditioning, and fancier options—in brief, the birth of the SUV.

When gas prices rise significantly, as they did again in 2007 and 2008, growing numbers of commuters form car pools, effectively multiplying per capita mpg; when prices subsequently decline, though, the appeal of car pools recedes—and increases in fuel economy have the same effect. Buy a Prius and, suddenly, the idea of sharing your commute (and your audiobook) with some stranger seems both less necessary and less appealing—especially if your city or state gives hybrids access to so-called car-pool lanes, as some do. According to a 2011 *New York Times* article by Sabrina Tavernise and Robert Gebeloff, carpooling by workers has decreased by almost half since 1980—a "rebound" response to real and perceived declines in the cost of driving over the same period.

There are much broader effects as well: efficiency chain reactions. A good way to see them is to think about refrigerators, the very appliances that the World Economic Forum and Steven Chu have pointed to as efficiency role models for global reductions in energy use. The first refrigerator I remember is the one my parents owned when I was little. They acquired it when they bought their first house, in 1954, a year before I was born. It had a tiny, uninsulated freezer compartment, which seldom contained much more than a few aluminum ice trays and a burrowlike mantle of frost. (Frost-free freezers stay frost free by periodically heating their cooling elements—a trick that wasn't in use yet.) In the sixties, my parents bought a much-improved model—which presumably was more efficient, since the door closed tight, by means of a rubberized magnetic seal. But our power consumption didn't fall, because the old refrigerator didn't go out of service; it moved into our basement, where it remained plugged in for another twenty-five years—mostly as a warehouse for beverages and leftovers—and where it was soon joined by a stand-alone freezer. Also, in the eighties, my father added an ice maker to his bar, to supplement the one in the kitchen fridge.

This escalation of cooling capacity has occurred all over suburban America. The basement of a vacation house belonging to a friend of mine contains three semi-retired refrigerators, one of which has been modified to serve exclusively as a beer-keg cooler, and another of which has a freezer compartment that's used only to chill beer mugs. The recently remodeled kitchen of another friend has an enormous stand-alone refrigerator, an enormous stand-alone freezer, and a two-drawer under-the-counter minifridge for beverages. (And that's just the kitchen.) Several people I know own dedicated, glass-fronted wine refrigerators with two cooling zones, one for red and one for white. (These are not luxury appliances. Costco sells a one-zone, fifteen-bottle model for $179.99.) In 2010, a *Wall Street Journal* article described the home of a wealthy Californian who had installed a refrigerator (as well as a coffeemaker) in his closet.

Nor has the trend been confined to households. As the ability to efficiently and inexpensively chill things has grown, so have opportunities to buy chilled things—a potent positive-feedback loop. Gas stations now often have almost as much refrigerated shelf space as the grocery stores of my early childhood; the explosive growth in the consumption of bottled and canned beverages, including the rise of the twenty-ounce single-serving soft drink, is partly both a product of and a contributor to the spread of refrigeration; even mediocre hotel rooms now usually come with their own small fridge (which, typically, either is empty or—if it's a minibar—contains mainly things that don't need to be kept cold), as well as an ice maker and at least one refrigerated vending machine down the hall. The fact that I am able to eat foods from all over the world, twelve months of the year, is largely a consequence of the vastly expanded cooling capacity of farmers, truckers, shippers, wholesalers, distributors, retailers—and me.

The steadily declining cost of refrigeration has made eating much more interesting, both for people who like fresh produce and for people who like processed junk. It has also made almost all elements of food production more cost-effective and energy efficient: milk lasts longer if you don't have to keep it in a pail in your well. But there are environmental downsides, beyond the obvious one that most of the electricity that powers the world's refrigerators is generated by burning fossil fuels. James McWilliams, the author of *Just Food*, told me, "Refrigeration and packaging convey to the consumer a sense that what we buy will last longer than it does. Thus, we buy enough stuff to fill our capacious Sub-Zeros and, before we know it, a third of it is past its due date and we toss it." (The item that New York City residents most often throw away unused, according to a consultant to the Department of Sanitation, is vegetables.) Jonathan Bloom, who runs the Web site Wasted Food (www.wastedfood.com) and is the author of *American Wasteland*, told me that, during the three and a half decades since the mid-1970s—the period during which refrigerators became bigger yet cheaper and more efficient—per capita food waste in the United States has increased by half, so that we now throw away 40 percent of all the edible food we produce.* And when we throw away food we don't just throw away nutrients; we also throw away the energy we used in keeping it cold as we lost interest in it, as well as the energy that went into growing, harvesting, processing, transporting, and preparing it (assuming we got that far), along with its proportional share of our staggering national consumption of fertilizer, pesticides, irrigation water, packaging, and landfill capacity. According to a 2009 study, more than a quarter of U.S. freshwater use goes into producing food that is later discarded. And rotting food is the main source of the methane—an especially worrisome greenhouse gas—that leaks from landfills.

*A boarding school in my town reduced food waste by 23 percent—in a week!—by eliminating trays from its dining hall, making it harder for students to carry more than they could eat—that is, by making students less efficient at transporting food.

20

Rebound Creep

Efficiency improvements are not limited to energy. They take place throughout the global economy and push down costs at every level—from the mining of raw materials to the fabrication and transportation of finished goods to the frequency and intensity of actual use—and falling costs stimulate both manufacturers and consumers. (Coincidentally or not, the growth of American refrigerator volume has been paralleled by the growth of American body mass index.) Harry Saunders has observed that efficiency increases occur in all factors of production—capital, labor, materials—and that all such gains also increase energy use, as well as synergistically reinforcing one another. He cites as an example the primary-metals sector of the American economy—a sector that comprises companies which produce raw metals and alloys, mainly from ore and scrap. Between 1980 and 2000, he writes, this sector experienced "the aggressive introduction of electric arc furnaces for steel production that were highly efficient in the use of both capital and labor, in addition to energy" and this transformation contributed to average annual efficiency gains of 2.46 percent for capital, 3.30 percent for labor, 2.90 percent for energy, and 0.53 percent for materials—and resulted in an overall "'all factor' energy rebound" of 172 percent, thus more than negating any energy "savings" from efficiency gains.

"By this analysis," Saunders concludes, "the increased efficiency of other factors not only increased energy consumption in this sector, but created significant backfire—a rebound in energy demand greater than 100 percent. The consequence of this phenomenon is rather profound. The problem is not so much that efficiency gains targeted at energy often also improve the efficiency of other factors (a feature of energy efficiency that analysts such as Amory Lovins cite as a key ancillary benefit); the real problem is that technology gains, considered together, increase energy consumption. Without these gains, energy consumption would be lower. Analytically, this makes 'teasing out' energy-specific rebound effects extra challenging. But the larger problem is that from a climate perspective technology gains generally are a culprit in increasing energy use."

In any context other than energy, such an observation would be uncontroversial. No economist would argue, for example, that making manufacturing plants more efficient causes total manufacturing to shrink. Len Brookes told me, "If you take all of the resources available to you and succeed in raising the productivity of one of them, in relation to the others, then that particular one tends to have a higher level of employment in the economy. Now, this argument you see most clearly with labor. If you persuade the workers that they should increase their productivity, all of past history shows that this increases their employability. And there's no reason that that should be any different for energy. If you increase the productivity of energy, then this increases its employment level in the economy."

Efficiency-related increases in one category, furthermore, spill into others. Refrigerators are the fraternal twins of air conditioners, which use the same energy-hungry compressor technology to force heat to do something that nature doesn't want it to. Because of that technological relationship, innovations that push down the cost of refrigeration also push down the cost of air-conditioning—thereby increasing its attractiveness to consumers—and vice versa. When I was a child, cold air was a far greater luxury than cold groceries. My parents' first house—like 88 percent of all American homes in 1960—didn't have even a window air conditioner when they bought it, in 1954, although they broke down and got a unit for their bedroom during a heat wave that summer, when my mom was pregnant with me; their second house had central air-conditioning, but running it seemed so expensive to my father that, for years, he could seldom be persuaded to turn it on, even at the height of a Kansas City summer, when the air was so humid that it felt like a swimmable liquid. Then he replaced our ancient Carrier unit with a modern one, which consumed less electricity, and our house, like most American houses, evolved rapidly from being essentially un-air-conditioned to being air-conditioned all summer long.

Modern air conditioners, like modern refrigerators, are vastly more energy efficient than their mid-twentieth-century predecessors—in both cases, partly because of tighter standards established by the Department of

Energy. But that efficiency has driven down their cost of operation, and manufacturing efficiencies and other productivity gains have driven down the cost of production, and those trends acting together have fueled market expansion, and the resulting economic growth has increased our wealth and therefore our ability to buy more. One consequence is that the ownership percentage of 1960 has now flipped: by 2005, according to the Energy Information Administration, 84 percent of all U.S. homes had air-conditioning, and most of it was central. Stan Cox, the author of *Losing Our Cool*, told me that, between 1993 and 2005, "The energy efficiency of residential air-conditioning equipment improved 28 percent, but energy consumption for AC by the average air-conditioned household rose 37 percent." That increase has been exacerbated by the fact that once people have air-conditioning they forget how to keep cool without it. My grandparents lived without air-conditioning in a hot part of the country but still managed to survive virtually a century apiece—and even in August my grandfather never took off his tie. They controlled summer temperatures by placing awnings over their windows, opening their windows and curtains at night and closing them in the morning, and, on especially hot nights, running a big whole-house fan that looked like a propeller salvaged from the *Titanic*. When I spent the night at their house during the summer, I would sleep on top of my covers with my head by the open window at the foot of my bed, and the basement zephyr would carry me off to sleep. I liked lying with my head by the window, because that way I could directly observe the many personal problems of my grandparents' neighbors, a large dysfunctional family.

In most of the United States today, such low-tech cooling techniques have essentially disappeared, to such an extent that when our houses feel hot to us we don't even bother to draw the curtains, but instead reach for air-conditioner controls. One result, Cox has observed, is that we now use roughly as much electricity to cool buildings as we did for all purposes in 1955. Another is that a room that used to be a standard feature of houses in many parts of the country, the screened porch, has become far less common. (If you don't have air-conditioning, the screened porch is usually the coolest room in your house: it's where you go in the evening to beat the heat. But once you do have air-conditioning the screened porch immediately becomes the hottest room in your house, and often seems unbearable by comparison. Once people who own old houses have added air-conditioning, they often enclose, air-condition, and heat their porches, so that they can continue to use them, year-round, without feeling uncomfortable. Energy use begets energy use.)

As *Losing Our Cool* clearly shows, similar effects permeate the economy. The same technological gains that have propelled the growth of U.S. residential and commercial cooling have helped turn automobile air conditioners, which barely existed in the 1950s, into standard equipment on even the least luxurious vehicles. (Similarly: power windows. In the United States, hand-cranked car windows are now almost as rare as hand-cranked car engines.) According to the National Renewable Energy Laboratory, running a midsized car's air conditioner increases fuel consumption by more than 20 percent—but the effects reach far beyond automotive cooling. Owning a comfortable car makes people willing to drive more miles and to endure commutes that would have seemed intolerable just a generation or two ago, thereby adding impetus to suburban sprawl and further reducing the appeal of (and demand for) public transit. Furthermore, access to cooled air is self-reinforcing: to someone who works in an air-conditioned office and drives an air-conditioned car, living in an un-air-conditioned house quickly becomes intolerable. A resident of Las Vegas once described cars to me as "devices for transporting air-conditioning between buildings."* In 1992, Gwyn Prins, a Cambridge University anthropologist, wrote that "physical addiction to air-conditioned air is the most pervasive and least noticed epidemic in modern America." One sign of our dependence is the declining significance of seasonal clothing. The year-round business suit is a product of air-conditioning; so are the tweed sports jackets worn by Hollywood executives in mid-July.

Brookes told me, "Critics will say there's a limit to how much you can backslide in your house. But you have to point out to them that they're not taking into account the fact that, if you really do make it cheaper to get your home heating or central air-conditioning, then the demand for a better standard of home heating or air-conditioning goes further down the income spectrum." In less than half a century, it has become possible for Americans of even

---

* It's often argued that a car with the air conditioner running uses less fuel at highway speeds than a car with the air conditioner turned off and the windows rolled down, because closing the windows makes the car more aerodynamic and therefore more fuel efficient. This may or may not be true, but it ignores the main environmental impacts of automotive air-conditioning, which have to do with its effect on vehicle miles traveled and on overall dependence on cooled air.

relatively modest means to spend entire summer days without passing more than a few moments in air that hasn't been artificially chilled—from home to car to work to shopping mall to home. (And although, as Lovins points out, we Americans don't use our more efficient furnaces to heat our houses to "sauna temperatures" during the winter, we do now heat much more than twice as much living space per person as we did 1950.) These same forces have accelerated the spread of cooling technologies all over the world. According to Cox, between 1997 and 2007 the use of air conditioners tripled in China (where a third of the world's units are now manufactured, and where many air-conditioner purchases are subsidized by the government). In Dubai, hotel swimming pools are often chilled, rather than heated, to keep swimmers from feeling poached. In India, air-conditioning is projected to increase tenfold in metropolitan Mumbai.

To suggest a causal connection between increased efficiency and increased consumption in this way strikes many economists and others as entirely misguided. Michael A. Levi, who is the David. M. Rubenstein senior fellow for energy and the environment at the Council on

Foreign Relations, has written that "declining appliance prices have nothing to do with increased efficiency—in fact, everything else being equal, increased efficiency leads to higher appliance prices (because the equipment seller captures part of the energy cost savings)." Furthermore, he writes, "my guess is that the spread of air conditioners (as well as cars and other such things) is driven mainly by the facts that people have more money to spend and that the devices are cheaper. The reduced cost of fueling them, I suspect, is a distant third."

This seems logical; but it's the kind of narrow, short-term, "bottom-up" analysis that Brookes and Saunders believe to be not only inadequate but misleading, since it focuses on specific end uses by consumers rather than on long-term macroeconomic effects. It also begs the question of where people get "more money to spend" and what makes devices cheaper, since even efficiency mavens treat increased efficiency as a form of income: it's "the lunch you're paid to eat." (Even so, Saunders has shown, in the United States between 1987 and 2002, household energy use rose in every income category and was therefore driven by more than income increases alone.) Efficiency improvements, furthermore, produce what Jesse Jenkins, who is the director of energy and climate policy for the Breakthrough Institute, has described as "frontier effects"—which, he writes, "result when efficiency gains unlock whole new areas of the production possibility frontier, leading to potentially vast new markets, or even whole new industries for energy services."

Imagine that you are an electronics engineer at Bell Labs in the 1940s. You feel frustrated by the large size, cost, and energy requirements of vacuum tubes, and you wish you had access to something that performed the same functions but was smaller, cheaper, and more energy efficient. Then, in 1947, your colleague William Shockley and his team develop the transistor, which answers all those needs, and within a relatively short time the vacuum tube is on its way to becoming obsolete. But the transition from tubes to transistors doesn't result only in the more-efficient redesign of electronic devices that existed in 1947: smaller, less energy-hungry radios; television sets that sit on narrower tables and don't need to "warm up"; computers that are indistinguishable from Second World War–era computers except that they consume less power and fit into smaller rooms. Instead, entire new categories of energy consumption arise almost instantaneously—frontier effects—and all that new consumption accelerates and amplifies as transistors become still smaller, cheaper, and more efficient. Viewed solely in the context of 1947, the transistor is a brilliant breakthrough in efficiency, dematerialization, and decarbonization: a portal to a low-energy future. But from the vantage point of the early twenty-first century, six and a half decades after Shockley's innovation, we can easily see that its real impact has been utterly different. Modern transistors are almost infinitely smaller and more energy efficient than their mid-twentieth-century equivalents (since they're now etched onto computer chips and individually require only infinitesimal amounts of power)—but my house contains billions of them, and most of them perform functions that no one in 1947 could have anticipated. As a consequence, my electronics-related consumption of energy and other resources has soared, not fallen—and so has the world's. And, despite what Amory Lovins and other efficiency mavens have repeatedly claimed, the drop in unit energy consumption and the rise in global energy consumption are not unrelated.

Frontier effects can work in both directions, since new markets and new industries often displace or even obliterate inefficient old ones. The free navigation app on my Droid X phone has made my year-old Garmin GPS device seem just about useless; the video camera on the iPhone killed the Flip. But if you widen your point of view, so that it takes in the entire economy, you can easily see that the overall trend, historically and globally, has always been in the direction of *more*. Efficiency gains of all kinds have enabled modern workers to accomplish in minutes tasks that used to require hours, even weeks. But that doesn't mean we call it quits after a few minutes, put up our feet, and spend the rest of the day twiddling our thumbs. We keep working and earning and spending and consuming—and we have the energy consumption, carbon output, and three-car garages to prove it.

Economic growth, by any definition, is the cumulative result of a vast and complexly interconnected web of factors, including productivity gains and efficiency improvements. And it's not a force of evil, since it's responsible for virtually all the tremendous comforts of modern life, including the innumerable reasons I'm grateful I'm alive today and not a hundred years ago. But economic growth, no matter how it arises, has environmental consequences, too. If, in Vaclav Smil's memorable phrase, energy flows are "the only real currency in the biosphere," the ultimate source of our riches is clear enough. The issue is whether we have the moral courage and political will to try to bend it in a different direction.

21

The Importance of Less

Not all efficiency gains stimulate consumption. When New York City began requiring more efficient plumbing fixtures, in the 1970s, water use in the city fell significantly and stayed down. The reason is that New Yorkers were already using as much water as they wanted to, since they weren't paying anything for it. (Water use in the city in those days wasn't metered.) And there really wasn't anything else that New Yorkers could have done with more water, since they didn't have yards or gardens or swimming pools or rooftop fountains and had no possibility of adding them. There was no "rebound": New Yorkers didn't start going to the bathroom more often just because their toilets were now using half as much water as before, and the city's population didn't rise, because the supply of water hadn't been holding it back. And, because city residents weren't directly paying for the water they did use, the drop in their consumption didn't affect them like an economic windfall: it didn't give them the means to increase their consumption in some other area.

But, Amory Lovins notwithstanding, most efficiency gains don't act like that, because, in most applications, we have little difficulty in finding ways to turn our savings into additional consumption. Even if we reach a point where our own activity in some category seems "saturated"—say, because we've fully paved over the country and can't think of anywhere else to drive, or because we've heated every room in our house to the maximum temperature we can stand, already leave windows open at night during the winter, and aren't interested in owning a bigger house—any additional efficiency gains will still serve to increase our wealth and will therefore give us the means to increase our consumption in other ways. In addition, the technology improvements that increase the efficiency of American machines don't belong exclusively to Americans. They spread to parts of the world where consumption still has lots of room to grow, and they stimulate it further.

Making U.S. cars less costly to operate makes other countries' cars less costly to operate, too—a bad thing, if the goal is to contain the direct and indirect environmental damage that is attributable, globally, to automobiles. The last thing the world needs is an inexpensive car that gets a hundred miles to the gallon, because once we have it there will no longer be a significant barrier, worldwide, to becoming a driver. And, as American history shows clearly, increasing the pool of drivers sets off a cascade of interconnected, seemingly irreversible environmental crises. More cars mean more roads, more roads mean more suburbs, and more suburbs mean more energy use and environmental damage in every category. And that goes for hybrids and electric cars every bit as much as it does for those powered by gasoline.

All such increases in energy-consuming activity can be considered manifestations of the Jevons paradox. Teasing out the precise macroeconomic impact of a particular efficiency improvement isn't just difficult, however; it's impossible, because the endlessly ramifying network of interconnections is too complex to yield to empirical, mathematics-based analysis. Most modern studies of energy rebound are "bottom-up" by necessity: it's only at the micro end of the economics spectrum that the number of mathematical variables can be kept manageable. (Jevons himself can be blamed for some of this difficulty; one of his contributions to economics was a greatly increased use of empirical, mathematics-based analysis.) But looking for rebound only in individual consumer goods, or in closely cropped economic snapshots, is as futile and misleading as trying to analyze the global climate with a single thermometer.

Lee Schipper told me, "In the end, the impact of rebound is small, in my view, for one very key reason: energy is a small share of the economy. If 60 percent of our economy were paying for energy, then anything that moved it down by 10 percent would liberate a huge amount of resources. Instead, it's between 6 and 8 percent for primary energy depending on exactly what country you're in." ("Primary energy" is the energy in oil, coal, wind, and other natural resources before we've used them to generate electricity or converted them into refined or synthetic fuels. Primary energy is the natural gas that runs the power plant, not the electricity the plant produces—a little like the difference between raw materials and retail goods.) Schipper's argument is that, because we can extract vastly more economic benefit from a ton of coal than nineteenth-century Britons did, efficiency gains now have much less power to stimulate consumption. But even if this is true, a very large fraction of the world's population today is in countries that are modernizing even faster than nineteenth-century Britain and converting fossil fuels into prosperity more aggressively than twentieth-century America. Besides, the fact that the percentage of global economic activity attributable to industrial energy production is small greatly understates energy's economic role. The logic misstep is apparent if you imagine eliminating primary energy from the world. If you do that, you don't end up losing "between 6 and 8 percent" of current economic activity, as Schipper's formulation would suggest; you lose almost everything we think of as modern life.

Blake Alcott, an ecological economist, has made a similar case in support of the existence of large-scale Jevons effects. Recently he told me, "If it is true that greater efficiency in using a resource means less consumption of it—as efficiency environmentalists say—then less efficiency would logically mean more consumption. But this yields a reductio ad absurdum: engines and smelters in James Watt's time, around 1800, were far less efficient than today's, but is it really imaginable that, had technology been frozen at that efficiency level, a greater population would now be using vastly more fossil fuel than we in fact do?" Contrary to the argument made by "decouplers"—whom I discussed in chapter 3—we aren't gradually reducing our dependence on energy; rather, we are finding ever more ingenious ways to leverage BTUs. Between 1984 and 2005, American electricity production grew by 66 percent—and it did so despite steady, economy-wide gains in energy efficiency. The increase was partly the result of population growth; but per capita energy consumption rose, too, and it did so even though energy use per dollar of GDP fell by roughly half. Besides, population growth itself can be a Jevons effect: the more efficient we become, the more people we can sustain; the more people we sustain, the more energy we consume.

In 1976, Amory Lovins, in a celebrated article in *Foreign Affairs* called "Energy Strategy: The Road Not Taken?" argued that the United States faced a choice between its current, environmentally perilous energy policy, which depended on the steadily increasing use of fossil fuels and nuclear power, and what he described as a "soft energy path," based on renewables, conservation, and efficiency. The conventional interpretation of our energy history since then is that America and most of the rest of the world have chosen the hard path over the soft, but in reality we've followed both. Nearly every energy-using device I own today is vastly more efficient than its 1976 equivalent: my house is better insulated; my furnace produces more heat from less oil; my windows are more weather-tight; my dishwasher and washing machine use less water, electricity, and detergent; and my car gets half again as many miles to the gallon despite being faster, heavier, less polluting, more mechanically reliable, and more equipped with fancy accessories. Yet my energy use and environmental impact have risen, because I have used my efficiency gains to leverage increases in my consumption, not to shrink it, and to satisfy wants that, forty years ago, I didn't know I had. Believing that we can address our energy and climate problems with efficiency gains and other "soft-path" strategies is like believing that homeowners can make their debt problems go away by increasing their charging limits on their credit cards. Lovins is undoubtedly correct when he says that we could live regally on little more than what we currently waste. But turning reduced waste into reduced consumption is a trick we haven't figured out. Paying the world to eat lunch, so far, hasn't caused the world to lose weight.

22

What Would a Truly Green Car Look Like?

The Ford Model T was manufactured between 1908 and 1927. According to the Ford Motor Company, its fuel economy ranged between thirteen and twenty-one miles per gallon. There are vehicles on the road today that do worse than that; have we really made so little progress in more than a hundred years?

But focusing on miles per gallon is the wrong way to assess the environmental impact of cars. Far more revealing is to consider the productivity of driving. Today, in contrast to the early 1900s, any American with a license can cheaply travel almost anywhere, in almost any weather, in extraordinary comfort, without leaving a paved surface; can drive for thousands of miles with no maintenance other than refueling; can navigate without consulting paper maps or asking for directions; can easily find gas, food, lodging, and just about anything else within a short distance of almost any road; and can order and eat meals without undoing a seat belt, ending a telephone conversation, pausing a recorded book, or turning off a ceiling-mounted DVD player. The car, for many Americans, has become a spa-like relaxation center, as well as a sort of mobile family room. Every suburban parent knows that the easiest way to have a frank one-on-one with a child is to take the child for a drive, since the child is strapped into a seat in more or less the posture of a psychotherapy patient and can't escape.

A modern driver, in other words, gets vastly more benefit from a gallon of gasoline—makes far more economical use of fuel—than any Model T owner ever did. But we have used those remarkable efficiency gains to increase our consumption, not to reduce it, and we now depend on our cars in ways that our grandparents and great-grandparents could never have imagined. Given that dependence, it shouldn't be surprising to us that our driving-related energy use has grown by mind-boggling amounts. U.S. consumption of motor gasoline has risen from about 11.5 million gallons per day in

1920, to 43 million in 1930, to 110 million in 1950, to 243 million in 1970, to 304 million in 1990, to approximately 390 million today. As always, the problem with efficiency gains is that we inevitably reinvest them in additional consumption. Paving roads reduces rolling friction, thereby boosting miles per gallon, but it also makes distant destinations seem closer, making it easier for us to drive longer distances and enabling us to live in new, sprawling, energy-gobbling subdivisions far from where we work and shop. And the effect is self-reinforcing, because living in those subdivisions further increases our dependence on cars, and so pushes up the number of miles we drive in all our other activities. When efficiency advocates say that automotive efficiency initiatives lose only 10 percent of their fuel savings to rebound, they make it clear that they're not looking at the real issue.

Steven Chu has said that drivers who buy more efficient cars can expect to save thousands of dollars in fuel costs; but, unless those drivers shred the money and add it to a compost heap, the environment is unlikely to come out ahead, since even if drivers don't spend their savings on more driving they will certainly spend them on goods or activities that involve energy consumption in some other form. The problem is exactly what Jevons said it was: the economical use of fuel is not equivalent to a diminished consumption.

Efficiency proponents often express incredulity at the idea that squeezing more work from less fuel could carry an environmental cost. Amory Lovins once wrote that, if Jevons's argument is correct, "we should mandate inefficient equipment to save energy." As Lovins intended, this seems laughably illogical—but is it? If the only motor vehicles available today were 1920 Model Ts, how many miles do you think you'd drive each year, and how far do you think you'd live from where you work? (If we all drove Model Ts, there would be no such thing as "NPR moments"—those times when a driver circles the block or sits in the driveway with the engine running in order to listen to the end of "A Prairie Home Companion.") No one is going to "mandate inefficient equipment," but unless we're willing to do the equivalent—by mandating costlier energy or finding other ways to dramatically reduce our total consumption—increased efficiency, as Jevons predicted, can only make our predicament worse.

Or maybe mandating inefficient equipment wouldn't be a terrible idea. During a talk I gave in New York in

2011, I described one possible vision of a green automobile: no air conditioner, no heater, no radio, unpadded seats, open passenger compartment, top speed of twenty-five miles an hour, fuel economy of five or ten miles a gallon. You'd be able to get your child to the emergency room, but you'd never run over to Walmart for a bag of potato chips, and you'd take public transportation to work. It's a given among environmentalists that Americans pay too little for gasoline, since the relative bargain pushes up our energy consumption by encouraging us to drive too much. Yet increasing fuel economy is the exact economic equivalent of *reducing* the price of gasoline, since doubling a car's miles per gallon has the same effect, on a driver's wallet, of halving the cost of fuel. One of the arguments made by the proponents of electricity- and methane-powered vehicles is that they will (in theory) substantially lower the cost of driving. How likely is that to weaken our infatuation with automobiles?

Decreasing consumption of fossil fuels is a pressing global need, for many reasons: climate change, dependence on undependable sources, habitat destruction, old-fashioned air and water pollution. The question is whether improving efficiency, rather than taking direct steps to constrain total consumption—the "cap" in "cap and trade"—can possibly bring about the desired result. Steven Chu told me that one of the appealing features of the efficiency discussions at the international Clean Energy Ministerial in 2010 was that they were never contentious. "It was the opposite," he said. "No one was debating about who's responsible, and there was no finger-pointing or trying to lay blame." This seems encouraging in one way, but dismaying in another. Given the known level of global disagreement about energy and climate matters, shouldn't there have been some angry table-banging?

Advocating efficiency is easy to do, because it involves no political risk—unlike backing measures that do call for sacrifice, such as increasing energy taxes, or putting a price on carbon, or capping consumption, or steadily rolling back total emissions, or investing heavily in utility-scale renewable-energy production, or confronting the deeply divisive issue of global energy equity, or radically redistributing the world's energy wealth. By comparison, improving efficiency carries no obvious cost. But efficiency initiatives make no sense, as an environmental strategy, unless they're preceded—and more than negated—by measures that force major cuts in total energy use. After all, improving efficiency isn't something we just invented. We've been doing it, globally, for centuries. It's how we made ourselves rich and, therefore, how we created the energy and climate problems that we're now trying to solve.

Herman E. Daly—an ecological economist and professor emeritus at the School of Public Policy of the University of Maryland—has written of the environmental necessity of imposing frugality (i.e., artificially increasing energy's scarcity through caps or taxes) before promoting efficiency (i.e., artificially increasing energy's abundance). He has written that "frugality first induces efficiency second; efficiency first dissipates itself by making frugality appear less necessary. Frugality keeps the economy at a sustainable scale; efficiency of allocation helps us live better at any scale, but does not help us set the scale itself." If we impose limits on our consumption of fossil fuels, advances in efficiency will enable us to live well with less damage; if we pursue efficiency alone, we will only make our problems worse.

Keynes didn't think much of *The Coal Question* (although, as with all of Jevons's work, he admired the prose). One of his objections had to do with Jevons's anxieties about the exhaustion of resources, a preoccupation that Keynes attributed at least partly to a character quirk. (Jevons, fearing an impending pulp shortage, hoarded so much brown packing paper that his descendants were still working through it years after his death.) But even if Jevons's apprehensions were in some measure neurotic they still led him to a counterintuitive insight that, a century and a half later, remains both provocative and timely. He died too soon to see the modern uses of oil and natural gas, and he obviously knew nothing of nuclear power. But his observations are less easy to dismiss than efficiency enthusiasts suggest. And worrying about the exhaustion of natural resources ought to seem less irrational to us today than it did to Keynes in 1936.

At the end of *The Coal Question*, Jevons concluded that Britain faced a choice between "brief but true greatness and longer continued mediocrity." His preference was for mediocrity, by which he meant something like what we mean by "sustainability." Our world is different from his, but most of the central arguments of his book still apply, adjusted for the passage of time, changed circumstances, and better statistical data. Steve Sorrell, a senior fellow at Sussex University and coeditor of a comprehensive recent book on rebound, called *Energy Efficiency and Sustainable Consumption*, told me, "I think the point may be that Jevons has yet to be disproved. It is rather hard to demonstrate the validity of his proposition, but certainly the historical evidence to date is wholly consistent with what he was arguing." That might be something to think about as we climb into our plug-in hybrids and continue our journey, with ever-increasing efficiency, down the road paved with good intentions.

23

Plentiful, Inexpensive Natural Gas Is Not an Environmental Solution

During the past half century, Americans have made so much real progress on a long list of once-pressing environmental problems—the Cuyahoga River doesn't burst into flames anymore; people not only swim but fish in Lake Erie and the Hudson River; Los Angeles enjoys occasional smog-free days—that we don't think about them nearly as much as we used to. And concerns about climate have pushed other environmental worries down the national anxiety list in the minds of many environmentalists. But most of our old problems are still with us, and—as is often the case in human affairs—they have sometimes been made worse by our efforts to address other environmental issues, however well-meaning those efforts may have been.

An obvious example is natural gas. Just a few years ago, U.S. gas reserves were thought to be relatively modest, and we were major importers from Canada. Since then, estimates of U.S. reserves have soared, and the price of gas has plunged. The reason is the rapid growth of an old extraction technique called hydraulic fracturing, or hydrofracking, which involves pumping water and other materials at high pressure into subterranean formations, shattering the rock and releasing gas (and oil) that is otherwise irretrievable. This development is generally presented as an environmental gain. Gas has the lowest carbon content of the fossil fuels, and now that it is suddenly more plentiful there's a clear economic incentive to replace coal-fired power plants with gas-fired ones, which run cleaner, and to replace furnaces powered by oil, and to convert vehicles that run on gasoline or diesel.

But there are complications. The main one is that gas is still a fossil fuel, and burning it still adds large quantities of carbon to the atmosphere. (Natural gas poses an even bigger climate challenge when it doesn't burn. When released into the atmosphere without combustion, it's a more destructive greenhouse gas than carbon dioxide, and undetected leaks from storage tanks, pipelines, and wells are a major contributor to atmospheric carbon levels—a fact that reduces its supposed climate advantage over other fossil fuels.) The natural gas industry, furthermore, invariably overstates the environmental gain from switching to gas, since even replacing old-technology coal plants with current-technology coal plants would represent a significant emissions improvement. In addition, hydrofracking does environmental damage unrelated to climate, most notably through the contamination of aquifers, wells, and watersheds. The water that's pumped into shale formations during hydrofracking includes chemical additives, and even when much of that water is recovered and "recycled" it often eventually ends up in municipal water treatment plants that were not designed to handle it. Hydrofracking may also trigger low-level seismic activity, as the drilling of deep geothermal wells is now also believed to do.

The drop in the price of natural gas has also added economic impetus to one of the world's most environmentally disastrous fuel-extraction technologies: the production of petroleum from so-called tar sands, the largest known deposits of which are in three vast formations in Alberta, Canada. To oversimplify: extracting liquid oil from tar sands is accomplished by heating, and the heating is done by burning natural gas. (The process also consumes huge volumes of water, as well as leaving behind permanently ruined landscapes.) Producing oil from tar sands is expensive, even if you don't count the cost of such "externalities" as ecosystem destruction, and an abundance of low-cost natural gas makes it cheaper. (The growing economic attractiveness of tar-sands oil is therefore partly a consequence of the greatly increased efficiency of modern techniques for extracting natural gas—a Jevons effect.) Processing tar sands consumes gas at a rate of as much as 1,200 cubic feet per barrel of recovered petroleum—and then, of course, we burn the petroleum, too. In 2010, Canada's National Energy Board estimated that by 2015 the tar-sands operation in Alberta would be burning 2.1 billion cubic feet of gas per day.

Perhaps most significant of all, the drop in the price of natural gas caused by the sudden expansion of its accessible supply has gutted any rational large-scale market for wind, solar, and other renewables. The low cost of coal has long been one of the main impediments to the development of nonfossil energy sources; now it's been joined by the low cost of natural gas. The entrepreneur T. Boone Pickens spent millions of dollars lobbying for a major wind-farm proposal of his—part of the "Pickens Plan," the stated goal of which was to end U.S. dependence on foreign energy—but dropped it after the low price of gas made wind uneconomical, even with its considerable subsidies. Pickens isn't necessarily a green role model, but his abandonment of his wind project indicates the challenge created by an abundance of cheap gas.

A further, more complicated problem is that gas, unlike coal, can be used directly as a fuel for motor vehicles. That means that, over the long term, the cheapness and ready availability of natural gas could reduce the environmental benefits of any increases in the price of oil, by making it easier for drivers to stay behind the wheel. And once enough vehicles are running on natural gas, producing natural gas from coal becomes attractive. One solution to these difficulties would be to impose a compensatory tax on natural gas in affluent countries. But the likelihood of that happening is probably less than zero. In fact, in the United States most of the legislative discussion of natural gas has concerned creating additional *subsidies* to encourage people to burn more.

Enthusiasts often speak of natural gas as a "bridge"

fuel—a methadone-like intermediate step between coal and an emission-free future. The idea is that an abundance of low-cost natural gas will give us the time we need to develop renewable, non-carbon substitutes, and will even (somehow) stimulate investment in them. But having access to a large supply of a low-cost, versatile resource does not stimulate investment in high-cost, problematic replacements. What natural gas is most likely a bridge to, eventually, is coal. (As gas gets more expensive again, coal will begin to look more interesting than it does at the moment.) And it's a mistake to think that a shrinking American market for coal is the same thing as a shrinking global market. Australia finances its occasionally impressive national interest in carbon reduction in part by exporting huge quantities of Australian coal to China.

The rebranding of natural gas as a green fuel is a remarkable achievement in marketing; at the 2011 Aspen Ideas Festival, I heard one speaker refer to gas, indirectly, as a non–fossil fuel. (Gas is the new wind!) But Carbon Lite is not the same thing as carbon-free. And non-climate environmental issues have not gone away.

24

Cheap, Efficient Lighting Is Not an Environmental Solution

Lighting fixtures have become so inexpensive to acquire and operate that we install them recklessly and in profusion—a seldom-mentioned downside to increases in lighting efficiency. In 2008, I flew into Atlanta late on a clear spring night and was struck, as I looked down during the plane's descent, by the fact that the largest visible terrestrial features were vast, empty parking lots, most of them illuminated so brightly, even at midnight, that you could have delivered babies in them. To see one source of the problem all you need to do is walk through the outdoor lighting section at Home Depot, where a five-hundred-watt halogen "wall pack" can be had for less than thirty dollars. Making lighting more energy efficient also makes it less costly, and that encourages us to use it carelessly.

The environmental consequences aren't all obvious. Unnecessary lighting fixtures and practices don't just squander energy and other resources; they also wreak havoc on ecosystems. In Florida, artificial lights have had a disastrous impact on sea-turtle populations. During the summer and early fall, hatchlings, which emerge primarily at night from nests on Florida beaches, are often fatally attracted to streetlights, house lights, and other sources of unshielded artificial illumination, and they die after being drawn onto roads or into open areas, where they are easily attacked by predators. The problem is that newborn sea turtles instinctively move toward the brightest part of the horizon—which, for millions of years, would have been not shopping malls and beach houses but the night sky over the open sea. Migrating birds can be fatally "captured" by outdoor fixtures, a fact that was made obvious a half century ago, when early versions of a common meteorological device called a ceilometer—which used a powerful vertical light beam to measure cloud ceilings—sometimes killed thousands of migrating birds in a single night. Artificial light can be especially lethal to insects. Gerhard Eisenbeis, a German entomologist, has written that outdoor installations can have a "vacuum cleaner" effect on insect populations, causing large numbers to be "sucked out of habitat." An earlier German study showed that new, brightly lit gas stations initially attracted large numbers of insects, but that the numbers fell rapidly after two years, presumably because local populations had been decimated. One of the several ways in which light fixtures kill insects is by disorienting them temporally, causing them to rest on the ground or in vegetation when they should be feeding or mating. And as insect populations fall so do the populations of the creatures that depend on insects for food, and of the creatures that depend on *them*. The modern American environmental movement was inspired, in 1962, by the publication of Rachel Carson's *Silent Spring*, which described the environmental impacts of pesticides; the global spread of inexpensive, highly efficient outdoor lighting has arguably been as ruinous—and it's more insidious, because the impact isn't obvious. (For clear depictions of the scale of this problem, type in "earth at night" at Google.)

The twenty-four-hour day/night cycle, which is also known as the circadian clock, influences physiological processes in virtually all living things. Pervasive artificial illumination has existed for such a brief period that not even the species that invented it has had time to adapt, biologically or otherwise. The most widely discussed human malady related to the disturbance of circadian rhythms is jet lag, but there are others. Richard Stevens, a cancer epidemiologist at the University of Connecticut Health Center in Farmington, has suggested a link between cancer and the "circadian disruption" of hormones caused by artificial lighting. Early in his career, Stevens was one of many researchers struck by the markedly high incidence of breast cancer among women in the industrialized world, in comparison with those in developing countries, and he at first supported the most common early hypothesis, which was that the cause must be dietary. Yet repeated studies found no clear link to food. In the early eighties, Stevens told me recently, "I literally woke up in the middle of the night—there was a street lamp outside the window, and it was so bright that I could almost read in my bedroom—and I thought, Could it be that?" A few years later, he persuaded the authors of the Nurses' Health Study, one of the largest and most rigorous investigations of women's medical issues ever undertaken, to add questions about nighttime employment, and the study subsequently revealed a strong association between working the night shift and an increased risk of breast cancer. Eva Schernhammer, of the Harvard Medical School, and Karl Schulmeister, an Austrian physicist, analyzed the work-shift data from the Nurses' Study several years ago and wrote, "We hypothesize that the potential primary culprit for this observed association is the lack of melatonin, a cancer-protective agent whose production is severely diminished in people exposed to light at night."

Nighttime lighting has seldom been a priority of environmentalists—one of whom described it to me as a "soft" issue. But the impacts are enormous, and they are all the more dangerous for being imperceptible. They also prove that "rebound" can take forms that have little or nothing to do with energy use.

25

Using Water More Efficiently Will Not Solve the World's Steadily Worsening Water Problems

An earthen causeway connects Pyramid Island—near the southwestern end of Lake Mead, twenty miles from the Las Vegas Strip—with the Boulder Harbor boat launch facility on Lakeshore Road. Two cantilevered piers extend like wings from the causeway's sides. There used to be a No Fishing sign at the end of one of the piers, but it served no purpose beyond stating the obvious: the lake's volume has shrunk by nearly 60 percent since 1998, and the piers overhang dry land. Some parts of the shoreline have receded by a half mile. A section of the lake to the south of the causeway used to be reserved for scuba diving; today, you can explore it in hiking boots.

Lake Mead was created during the Great Depression

by the construction of the Hoover Dam, which spans the Colorado River. Most people assume that its unnerving shrinkage is a result of the rapid growth of metropolitan Las Vegas—which depends on the lake for water and has quadrupled in population during the past two decades—but that's not the case. Colorado River water is apportioned according to a 1922 agreement, which annually allocates 15 million acre-feet among seven states. (An acre-foot equals a little less than a third of a million gallons.) Because Las Vegas was barely even a hick town when the agreement was signed, Nevada's share is tiny: just 300,000 acre-feet a year, or 2 percent of the total U.S. draw. (By comparison, California and Colorado, between them, take about 2.5 trillion gallons a year, or roughly half the total.) This limit has forced southern Nevada to leverage its allotment by adopting some of the most stringent conservation regulations in the United States, under the direction of the Southern Nevada Water Authority, a regional governing body. Every gallon that goes down a drain indoors in Las Vegas is treated and then either reused or returned to the lake, thereby earning a "return flow credit," and homeowners can be fined for doing things like watering their gardens on the wrong days of the week or allowing runoff from a sprinkler to flow onto a sidewalk. The SNWA also pays cash rebates to water customers who convert turfed areas into "xeriscapes." (The word comes from the Greek *xeros*, meaning "dry.") These efforts have been so effective that Nevada currently uses only about 265,000 acre-feet a year, or less than 90 percent of its allotment.

Yet Las Vegas's experience with water demonstrates the challenge, as well as the promise, of conservation and recycling as strategies for managing the world's freshwater resources. The city has cut its per capita water use to well below the U.S. average—but those same reductions helped to fuel its transformation into one of the most sprawling metropolitan areas in the country, by enabling the region to support a far larger population than would have been conceivable otherwise. When I visited the Pyramid Island causeway, I was accompanied by Nicole Lise, who works in the public information office of the SNWA. Lise first saw the city in the 1970s, on a vacation with her parents. The area's explosive population growth hadn't begun yet, and her family's first reaction was to wonder where all the hotel and casino employees went when their shifts ended. "We thought maybe they lived in the hotels," Lise told me, "because we didn't see any houses." No one visiting Las Vegas today would suffer the same misapprehension. The subdivisions seem to extend all the way to the mountains, and in every direction.

Water conservation efforts like southern Nevada's can have other negative consequences that are just as easy to overlook. The SNWA's turf replacement program requires that converted areas be planted with replacement vegetation and in such a way that the canopies of the plantings, at maturity, will provide ground coverage of at least 50 percent. The purpose of the rule is to ameliorate Las Vegas's airborne-dust problem, which is sporadically severe, and to reduce the ground's absorption and reradiation of solar energy—the so-called heat island effect, which can raise air temperatures and even alter local weather patterns. Pat Mulroy, the SNWA's executive director, told me, "The old mantra for desert landscaping in southern Nevada, when I first started, was a cactus, a rock, and a dead-cow skull. What we want is desert landscaping that has a single root system, but can spread all over the ground. That keeps the temperature of the rocks down—which is necessary because in the summer it gets pretty hot around here." But there are downsides. The main one is that the nonturf plantings have to be irrigated, too. Every plant I saw in a turf-removal area, other than weeds, had its own emitter, a small, black plastic nozzle that was feeding it controlled amounts of water. (One unanticipated difficulty with the nozzles is that they appeal to coyotes, which treat them the way my dog treats rawhide chews.)

Dale Devitt, a biology professor at the University of Nevada at Las Vegas, told me, "The popular idea is that if you remove turfgrass you're going to save unbelievable amounts of water, but the reality is that there are tradeoffs. Removing the turfgrass is one thing, but if you don't control what goes back in, and just plant trees instead, within a period of time there's no savings at all. We've demonstrated, for example, that one mature oak tree requires as much water as 1,600 square feet of low-fertility Bermuda grass. People will sometimes remove turfgrass but leave fifty-foot-tall trees behind, without realizing that the trees were totally dependent on the irrigation that the turfgrass was receiving." Devitt continued, "When you talk about water savings in a landscape, the big savings don't come so much from changing what you're growing. The big savings come from reducing the size of the landscape."

This is really the rebound problem all over again. What Devitt is saying is that the big savings don't come from using water less inefficiently; the big changes come from using less water, period.

And there are other challenges. Processed wastewater inevitably contains things that potable freshwater doesn't, primarily salts. This problem is acute in Nevada, because the Colorado River is high in salts to begin with and recycling nearly doubles their concentration, with the result that irrigating grass or other vegetation entirely with recycled water does the equivalent of applying ten or eleven tons of salt per acre per year. The salt harms the vegetation directly; it also creates a number of more subtle problems, by gradually clogging air and water channels in the soil, reducing the ability of roots to absorb nutrients, promoting previously unfamiliar plant diseases, and inducing the sprinkler-valve equivalent of atherosclerosis. "Salinity is one of those things you won't notice at first," Devitt said, "because the changes are very subtle, especially when you're growing salt-tolerant grasses like Bermuda grass. But at some point you cross a threshold, and it's downhill from there."

What appear at the time to be valuable environmental breakthroughs often turn out to be long-term disasters in the making. In the 1960s and '70s, the so-called Green Revolution transformed food production in India and a number of the world's poorest countries by improving irrigation systems, extending access to synthetic fertilizers and pesticides, and introducing high-yielding varieties of staple grains (most of which depend heavily on irrigation, synthetic fertilizers, and pesticides). Those advances transformed Third World food production and turned some poor countries from reservoirs of malnutrition into net food exporters. But they also helped to create what have turned out to be unsustainable environmental pressures, by draining aquifers, exhausting soil, and allowing populations to soar. In Syria, large areas of formerly arable land have turned to desert, and the country's freshwater stores have fallen by more than half since 2002. Such problems are not confined to developing countries. Farmland in some parts of California has subsided by more than thirty feet since the mid-twentieth century, as underlying aquifers have been depleted. Remedies often turn out to be worse than the problems they were intended to address. An irrigation technique that once seemed environmentally enlightened—repeatedly capturing irrigation runoff and reapplying it to the same fields—rapidly raises the soil concentration of salts and other undesirable substances and can eventually render the fields permanently unsuitable for growing anything.

When I asked Pat Mulroy whether metropolitan Las

Vegas might not ameliorate some of its water problems by taking steps to cap its population growth and halt its horizontal spread across the valley, she said, "Please share with me how you're going to do that. Under our constitution everyone who owns private land has a right to develop it to its highest and best use, so controls on that end have to come from the land-use side. What I have said to this community is, yes, you can continue growing, but you cannot do it the way you have in the past. You have limited water resources and you live in a fragile environment, so you're going to have to plan development that is much friendlier with outside water use."

Las Vegas's outward expansion has stalled for the time being—but as a result of the national mortgage meltdown and recession, not of any transformation of land-use policy. And that fact, paradoxically, has complicated southern Nevada's ability to maintain its remarkable record of success in controlling regional water use. The SNWA depends for its budget on the substantial connection fees paid by the owners of new homes. When construction stops, the fees stop, too.

26

Burning Trash Is Not the Answer

An obvious strategy for reducing consumption of fossil fuels is to shift more energy production to noncarbon sources—ideally, ones that are at least theoretically renewable. But doing that is neither easy nor straightforward, and it never has been. Jevons, a century and a half ago, offered a good explanation of why "alternative" energy sources, such as wind, hydropower, and biofuels (in his day, mainly firewood and whale oil), could not compete with coal: coal had replaced *them*, on account of its vastly greater availability, portability, utility, productivity, convenience, and energy content. (Early British steam engines were sometimes used to pump water to turn waterwheels.) Great Britain embraced coal partly because the country didn't have unlimited access to alternatives: in its quest for fuel, it, like large parts of continental Europe, had cut down and burned the vast majority of its trees, even though its population and per capita energy needs were modest in comparison with those of the twenty-first century. (The settlers who founded the British colony at Jamestown, in 1607, included German glassmakers because the venture's financial backers were hoping they would be able to ship cheap glass back to Britain, where glassmaking was constrained by the cost and scarcity of firewood.)

In 2010, the *New York Times* (as part of a series called "Beyond Fossil Fuels") published an article, by Elisabeth Rosenthal, about the Swedish city Kristianstad, which has systematically reduced its consumption of fossil fuels—to such an extent, the article reported, that "the city and surrounding country, with a population of 80,000, essentially use no oil, natural gas or coal to heat homes and businesses, even during the long frigid winters." Kristianstad has done this primarily by shifting most of its heating from the distributed combustion of fossil fuels to the centralized combustion of biomass—mainly food waste, farm waste, and wood. It has also converted its municipal vehicles from petroleum to locally produced biogas and is hoping to eventually transform the area's private-vehicle fleet as well.

To anyone who is concerned about energy and climate, this story seems highly encouraging. It's not about a few determined back-to-the-earth types living in yurts; it's about a real, decent-sized modern city—and in an unfavorable climate. And Kristianstad's experience does prove that, with a modest amount of ingenuity, people can harvest energy from materials that might otherwise end up in landfills. But the story glosses over critical details, as such stories invariably do. Kristianstad's biofuels program depends on very specific local circumstances, and the city's success does not scale up into anything like a global or national strategy.

The critical details have to do with the sources of the waste materials that Kristianstad burns. Those materials do not come—as a careless reader might assume—from the residents of Kristianstad alone; they come from three substantial regional industries: agriculture, food processing, and wood processing. Those industries don't serve just Kristianstad, and, therefore, the combustible waste they produce is actually attributable to a much larger population, not all of which is even in Sweden. If Kristianstad's residents attempted to heat their homes and businesses with just their per capita share of the waste products of their own consumption, most of them would freeze long before spring had arrived—and you can prove that to yourself by imagining heating your own house through a New England winter by incinerating only the stuff you add to your compost heap and set out for the trash collector. This kind of energy generation, furthermore, depends on the continued abundance of a resource that we are otherwise committed to reducing: waste.

Sweden is a heavily forested country with a population that's small in relation to its land area, and it has a significant wood-products industry, so it's well positioned to harvest BTUs by burning industrial wood waste. But such material isn't really an untapped resource, because most of it, worldwide, already is burned to generate heat or power, usually for the same factories that produce it. (In the United States, most of the energy production attributable to biomass—in fact, the largest share of energy production attributable to "renewable" sources—is created and exploited in that way.) Besides, burning biomass in any form is a tough way to produce energy. As Jesse H. Ausubel of Rockefeller University wrote in 2007, fueling a single thousand-megawatt electric power plant (half a Hoover Dam) solely with renewably harvested plant material would require the entire output of more than 2,500 square kilometers—roughly 620,000 acres—of prime agricultural land maintained in perpetuity at maximum productivity, using significant inputs of water, fertilizer, pesticides, and energy. "Prime land has better uses, like feeding the hungry," Ausubel concluded. And preserving prime land by growing energy crops only on poor land, he wrote, would increase the acreage requirement by a factor of ten or twenty, while also increasing the risk of erosion and the need for destructive inputs. Nature, furthermore, places a low limit on the maximum potential energy yield of plants, even on prime land: "Because more biomass quickly hits the ceiling of watts per square metre, it can become more extensive but not cheaper." In other words, you can sow more acres, but the amount of energy you harvest from each acre can't rise beyond a fairly low level—a threshold imposed not by technology but by biology and physics. A 2005 U.S. Department of Energy report concluded, "Almost all of the arable land on Earth would need to be covered with the fastest-growing known energy crops, such as switchgrass, to produce the amount of energy currently consumed from fossil fuels annually." A global commitment to biofuels would be a global commitment to mass starvation. Even now, relatively small changes in U.S. demand for corn—the staple crop of the phenomenally ill-considered American investment in ethanol—can push global food prices to catastrophic levels.

The area I live in is heavily wooded. In fact, there are many places within my town's borders from which no sign of habitation is visible in any direction. But the woods aren't old. I've seen photographs and postcards from the late 1800s and early 1900s that show very few trees. One reason is that, in those days, much local land had been cleared for agriculture; another is that most of the old-growth trees, including the ones that were cut down to create crop fields and sheep pastures, had been used in construction and as fuel. Some of the harvested wood was burned to heat buildings; some was turned into charcoal, which was burned to smelt iron. To early American settlers, the trees that covered our continent appeared so inexhaustibly abundant that the term they adopted for wooden building materials was the word the British used (and still use) for heaps of stuff just lying around: *lumber*. Even so, eighteenth- and nineteenth-century Americans rapidly decimated vast stretches of former woodland, and, as eastern forests fell, the logging crews moved west. Paul K. Barten, a professor in the Department of Environmental Conservation at the University of Massachusetts at Amherst, has written that, during a roughly thirty-year period beginning in the late 1800s, when the U.S. population was less than a quarter of what it is today, loggers in Minnesota alone clear-cut a forest area twice the size of the Adirondacks.

Then coal overtook wood as an energy source, for the reasons Jevons described, and the forests began to grow back—not only in my town but all over New England and other parts of the rest of the country. And then oil overtook coal. One of my regular golf buddies remembers helping to dismantle our town's old coal bins, down at the lumberyard, when he was a teenager, in the 1960s. The bins were no longer needed because residents were now heating their houses with oil instead of coal and firewood, and the local train service—which operated between 1872 and 1948—had been driven out of business by a more convenient, cost-effective, and efficient form of transportation: cars and trucks.

Reforestation seems like a very good thing. And the world's progression from wood to coal to oil and natural gas provides a clear example of the steady "decarbonization" of modern energy production. But the overall result of that progression has been the creation of the energy and carbon problems we're struggling with now, not their gradual elimination. Replacing fossil fuels with renewable ones—in effect, reversing the energy evolution of the past century and a half, a period during which the population of the world has more than quintupled—would require an almost unimaginable decrease in global energy use, because, without deep reductions in our appetite for power, renewables would never be able to keep up. They couldn't keep up in the United States in the 1800s, when both total population and per capita energy use were a fraction of what they are today; they certainly won't be able to keep up in the twenty-first century, when we've grown accustomed to burning fossil fuels to do things like making our toothbrushes move back and forth.

27

When Solar Power Isn't Green

All renewable nonfossil energy presents similar challenges—including solar. The sun is the ultimate source of most of the energy we use, since fossil fuels are the transformed remains of eons-old plant materials, whose energy content arose through photosynthesis. But raw sunlight is highly diffuse, and that means that to harvest useful quantities you have to cover a lot of ground. The main reason we find fossil fuels so handy is that they contain solar energy in extremely concentrated form—the result of millions of years of accumulation and compaction and other forces that would be impossible to duplicate artificially without an earth-sized machine that had a molten, radioactive core. Certain devices are able to convert sunlight directly into electricity, and certain other devices are able to concentrate sunlight sufficiently to heat liquids to generate steam to turn turbines. But there's a limit to the maximum possible electricity yield per unit of collection area, and that limit helps to explain why, in the second decade of the twenty-first century, despite decades' worth of talk and promises, solar still accounts for only a minuscule fraction of global energy production.

In the popular imagination, an easy solution to our energy and climate problems would be to install photovoltaic panels on all our roofs, plug in our electric cars, and proceed as before—and at a cost savings, to boot. And you can easily find stories in newspapers and on Web sites about homeowners who now generate so much power with solar panels that they are able to sell surplus current to their utility—a practice known as net metering. But real life is nothing like that simple. For one thing, the ideal residence for rooftop electricity generation is an oversized single-family, single-story dwelling on a large, treeless lot in a desert—an optimal configuration in terms of solar exposure but a disastrous one in terms of sustainable human communities. (A tall apartment building in a dense city is a far greener living arrangement for human beings, even though it offers little potential for rational solar collection.) For another thing, the real economic considerations are almost always hidden under layers of hype and subsidies and tax breaks, as well as burdened by glossed-over ancillary investments in supporting infrastructure, backup generation, and storage. Most suburban owners of solar panels could put a sign in their front yard saying "My neighbors pay my electric bills!" They would be far less enthusiastic if their personal power consumption were truly limited to what they could pull in from their own roof. (U.S. fossil-fuel consumption is also heavily subsidized, and not only directly but also in complicated ways that can be virtually impossible to untangle. But hype is hype.)

The fundamental inadequacy of conventional thinking about photovoltaics was neatly captured in 2007 by Ozzie Zehner, now a visiting scholar at the University of California at Berkeley, in his PhD thesis at the University of Amsterdam. Zehner described opening an office for his architectural practice and exploring the possibility of adding solar panels to his roof:

I was dismayed to find that no matter how firmly I massaged the numbers, the solar cells would never yield a monetary payback even under the most favorable weather conditions. I was further shocked by the embodied environmental costs to produce the solar arrays, from the mining of minerals to the transportation and installation of the finished product. Cleaning and maintenance costs were also projected to be high because of the relative inaccessibility of the rooftop units. Furthermore, they rely on a thinly spread solar radiation from a sun that only shines half of the day, a cosmological constant showing no signs of improving. I was also alerted by concerns about global solar markets. While solar technology holds meaningful hope for developing regions that lack energy grids, most cells are sent to rich countries like Germany and Japan. Elevated demand for silicon fuelled by subsidies in cloudy Germany was artificially raising prices out of the reach of those who could benefit the most. It seemed solar energy was evolving as a first-world fetish, an expensive one at that!

No genuine, large-scale conversion from fossil fuels to solar or other renewables will be possible unless per-capita energy consumption in wealthy countries falls, and by dramatic amounts. The easiest way to see this is to look closely at an actual, off-the-grid community that relies heavily on solar. One is Natural Bridges National Monument, a national park in southeastern Utah. The facility is far from any power lines. Until a little over thirty years ago, its electricity was provided entirely by diesel generators; then, in 1980, the federal government converted it to photovoltaics, as a demonstration project. The system—which has been upgraded and modernized since then—consists of a one-acre collection field, which is ideally situated for maximum exposure to the sun; a fifty-kilowatt inverter, which turns the direct current produced by solar panels into the alternating current used by appliances; thirty-nine 1,200-pound lead-acid batteries, which provide two days' worth of backup storage; and a huge diesel-powered generator, which kicks in during rare extended sunless periods and which, in a typical year, handles about 10 percent of the park's electricity consumption. This system powers the visitor center, restrooms, outdoor lighting, and other loads, including the light fixtures and electric appliances in the tiny, prefabricated houses occupied by the small staff. But even though those houses are minimally equipped with power-using devices—and even though almost all heating and cooking in the park is done not with electricity but with liquefied petroleum gas, which is brought in by truck, along with fuel for staffers' vehicles—the park still had to cut its presolar electricity consumption by two-thirds (from 200,000 kilowatt-hours a year in the diesel era to 70,000 kilowatt-hours today) to make the system viable. This discouraging arithmetic is seldom mentioned in discussions of solar's future—a future in which, it's often suggested, the sun will not only replace our current sources of electricity but also power our cars.

The diffuseness of sunlight means that serious reliance on solar power in any form will require not only steep reductions in total energy consumption but also vast installations, covering huge tracts of land—something that most Americans (including many American environmentalists) have shown little appetite for so far. In early 2011, five licensed utility-scale solar-thermal projects in California alone were halted by lawsuits, most of which focused on likely environmental impacts. In the meantime, solar evangelists usually argue that even modest installations—like those on suburban roofs—at least keep the technology moving forward. But the possibilities for continued technological advances are not infinite, and a focus on small-scale applications can have unintended consequences—among them the encouragement of sprawling residential development (and the infrastructure required to serve them), and the extension of human habitation into places that are remote from existing power grids and that could not have been developed if distributed energy sources did not exist.

Another problem is that, once I've got my own rooftop photovoltaic panels (financed partly by my fellow taxpayers and utility customers), my investment in my system is fully sunk, from my point of view, and I have no incentive to conserve. Even if my panels generate a surplus that I sell to my utility (at a price that is likely to be a multiple of what the utility would have had to spend to generate the same number of kilowatt-hours on its own, even from solar), I'm merely reducing the cost of my own electricity use, with predictable results. Imagine buying a photovoltaic-powered air conditioner (now available as a stand-alone unit, with no need for a separate inverter). Leaving it running while I'm at work, so that my bedroom will be comfortable when I return, will seem "free" to me, since no charge will appear on my electric bill—so why would I turn it off? And, if millions of other consumers made the same decision, the overall effect would be of a major national investment in doing more of something we should actually be doing less of. Household solar users are highly susceptible to a belief that consumption itself has environmental value if the energy-using devices it depends on are "green"—the Prius Fallacy again.

A trivial example, but a real one, is those solar-powered LED garden lights you increasingly see in suburban yards. They stick into the ground like tent stakes and have a small photovoltaic panel and a battery. They harvest and store solar energy during the day and use it after dark to provide illumination. But the homeowners who buy them at Walmart seldom use them to replace existing, wired fixtures—mostly because the light level and quality is very low and the battery doesn't last all night. Instead, they put their new solar lights in places they wouldn't have thought were worth illuminating if they'd had to pay an electrician to run a cable under their lawn. Such lights, though their operation consumes no fossil fuel, add to the world's energy load. That is, they increase overall energy use, rather than reducing it—a problem with all growth in consumption, but one that's especially hard to avoid with any energy source (including efficiency gains) that consumers perceive as both green and cost free. In 2011, at the Asia Society in New York City, I moderated a panel discussion called "Greening the Chinese City." One of the speakers, a Chinese architect and urban planner, showed a slide of a new Chinese expressway, which was illuminated by photovoltaic-powered streetlamps. The architect cited the lights as an example of "clean energy" and implied that the presence of PV panels transformed the highway, and even driving itself, into an environmental asset.

This challenge is more acute—and more morally complex—in places where solar installations are, as it were, a gateway power source. In parts of Africa and Asia, for example, photovoltaic panels are now beginning to provide electricity to impoverished regions too isolated to be served by any conventional grid. Those panels have brought light and power to homes, businesses, schools, and hospitals and have dramatically improved the lives and future prospects of people who, until recently, had been forced to subsist without access to almost anything resembling modern conveniences. But those panels also increase the world's energy load and, therefore, its long-term demand for all fuels, including fossil fuels. And that means they intensify the world's climate and energy predicament, rather than alleviating it.

In 2011, a tailor in Bangladesh told a reporter from CNN, "We could not work much before we got the solar power. We had to stop work before sun set. But now we can work until ten in the night so it boosts my income." From any humanitarian point of view, this is a highly desirable result; there are more than 160 million people in Bangladesh, and nearly half of them still have no access to electricity. But increased income inevitably means increased consumption, and increased consumption means increased energy use in all forms, as well as expanded demand for consumer goods—eventually including the ultimate modern consumer good, the automobile. Photovoltaic panels can enable isolated populations to reduce their reliance on firewood, dried dung, kerosene, diesel, and other problematic fuels, and they raise standards of living—all desirable results. But, in the end, new power demand in any form begets new power demand in every form. As Ethan Goffman asked recently on the Web site of the journal *Sustainability: Science, Practice, & Policy*, "As these clean energy sources come online, won't a world of voraciously expanding consumption be tempted to use even more energy, continuing to tap fossil fuels to their maximum along with the new sources? Or else buying and consuming other goods made available by a new source of abundant energy? Isn't it human nature to push the environment beyond its limits no matter what the technology?"

That doesn't mean that Bangladeshis should be denied access to solar panels—or that wealthy countries should attempt to crush development in less fortunate populations or in countries, like Egypt and Libya, whose economies (and therefore energy-consumption levels) have been stunted for decades by autocratic leaders. But it does mean that, if we're serious about both climate change and global equity, the actual challenges we face are vastly greater—and will require far larger reductions in consumption by the world's more fortunate citizens—than the ones we currently treat as inconceivably huge.

28

Green or Not Green?

Hydroelectric generation is the second-largest existing source of renewable energy in the United States, after the combustion of industrial wood waste and other biomass, and it's by far the largest renewable source that doesn't involve setting things on fire. Hydro has many appealing features as an energy provider, chief among them the fact that flowing water is usually more dependable than either sunshine or wind. The main downside is that dams disrupt ecosystems. Another is that we've already put hydroelectric plants just about everywhere we could rationally put them, and no one is creating new rivers. It's possible to add relatively small-scale generating capacity to existing dams, a technology known as incremental hydropower. But the main trend has been in the other direction. In fact, environmentalists have more often fought to remove existing dams than to build new ones or to increase the output of functioning power plants. In 2010, the electric utility PacifiCorp agreed to remove four power-generating dams from the Klamath River in California and Oregon, beginning in 2020. Similar river-restoration campaigns are under way elsewhere, and pressure for removal will likely increase as growing numbers of existing dams—many of which are more than fifty years old—deteriorate to the point of requiring hugely expensive repairs. (The Association of State Dam Safety Officials has said that 4,400 of the country's 85,000 dams are in danger of failing.) Dams are also wholly dependent on the continued abundance of the water backed up behind them. The surface of Lake Mead, which has lost more than half its volume, is still above the power plant's intakes, but by nowhere near as much as it used to be. The immense reservoir behind China's Three Gorges Dam—the world's largest hydroelectric power plant, with a generating capacity roughly nine times that of the Hoover Dam—has had to be tapped heavily to address severe drought conditions, of which the dam itself may be a principal cause.

Hydroelectric power poses additional perplexity: we tend to think of it as "free." The North American cities with the lowest prices for electricity are almost all ones that generate much of their power hydroelectrically. This rate structure makes perfect sense economically: lower production costs equal lower prices. But from an environmental perspective cheap energy is a problem, no matter what the source. Electricity customers in Seattle and Vancouver pay less than a third of what I do in Connecticut. Where's their incentive to conserve? This is the Prius Fallacy in another form. In the global scheme of things, recklessly consuming "green" energy is no greener than recklessly burning coal. Access to cheap electricity doesn't solve environmental problems; it amplifies them, by relieving the economic pressures that, if left in place, would force consumers to make real, lasting, environmentally beneficial reductions in consumption. But good luck convincing green-thinking residents of Seattle, Washington, or Vancouver, British Columbia, that they don't pay enough for electricity, or that they should think of their power consumption no differently than if it were fueled by coal.

In 2011, heavy rains in the Pacific Northwest filled to overflowing the rivers that power most of the region's hydroelectric plants, pushing electricity production to record levels. This sounds environmentally valuable, but it actually wasn't. Because the region (like the United States as a whole) lacks an interconnected network of long-distance transmission lines, the surplus electricity couldn't be sold to other markets. In fact, some of the region's wind turbines—which, because of unusually strong, steady winds, were also generating power at near-maximum levels—had to be shut down to prevent the surplus from becoming unmanageable. The end result was to hold down regional electricity prices (thereby encouraging additional consumption) without displacing generation from plants that run on fossil fuels. And opposition to the construction of new long-distance high-voltage transmission lines is one critical force standing in the way of significant exploitation of most renewable-energy sources, since those sources are not uniformly distributed across the country. The opposition comes not just from potentially affected private-property owners but also from state governments and competing utilities. This is an extremely unsexy green problem that won't be resolved without federal intervention.

29

Flying a Kite

Saul Griffith—the young Australian engineer I described in the first chapter, who won awards for a method of manufacturing cheap eyeglasses—was born in Sydney in 1974. He has an athletic build and a scruffy beard. His hair, which is reddish brown, is usually an omnidirectional mess, and he often looks as though he got dressed from the bottom of the laundry pile. He warned me, when I met him, that he despises clichéd descriptions of inventors' childhoods—always taking things apart to find out how they work, then blowing them up or setting them on fire—but that's the kind of childhood he had. When he was six or seven, he spent most of a summer trying to replicate Batman's grappling hook and kept at it until he was able to hoist himself onto the roof of his house, and he built a helicopter that he powered by attaching fireworks to its rotors. One day when he was a little older, while lifting a glass of milk, he began to consider the complex neuromuscular choreography involved in getting the glass to his mouth and, overwhelmed by the improbability of coordinating so many disparate elements, spilled the milk. A similar reflection drove him from golf, which he played at a very high level as a teenager but largely abandoned, at the age of eighteen, after having a paralyzing vision of the mechanical preposterousness of the golf swing.

He received his PhD in 2004. His thesis concerned machines that assemble and replicate themselves, based on information contained in their own components—a concept known as programmable matter. Such machines function in a manner roughly analogous to the growth and reproduction of living things, which build themselves, cell by cell, in accordance with rudimentary instructions contained within each nucleus. Griffith created puck-sized plastic components that could fasten themselves to one another in specific ways, and then he set them in motion on an air-hockey table. Random collisions on the table, over periods of hours, caused the components to join together and to correct their own assembly errors.

After MIT, Griffith moved to California and, with a half dozen friends and former fellow students, attempted to reproduce the creative environment they had known in Cambridge. They founded an independent research-and-development company, which they called Squid Labs. ("I think we chose the name because squids have good eyesight and large brains and are good at solving puzzles.") They rented a warehouse, adopted the slogan "We're not a think tank, we're a do tank," and acquired much of their laboratory equipment for little or nothing, through Craigslist. To save money, Griffith and his future wife, Arwen O'Reilly, lived in the warehouse's attic. (O'Reilly's father is Tim O'Reilly, the founder and CEO of O'Reilly Media. He was the creator of the world's first commercial Web site and the coiner, in 2004, of the term "Web 2.0.")

Squid Labs quickly became very productive. Griffith and his colleagues worked on an old MIT invention of his, electronic rope, which contains sensors that detect changes in load, and in 2005 it was selected by *Time* as one of the twenty-five best inventions of the year. They built a hand-operated power supply for the so-called hundred-dollar computer, on a grant from the nonprofit organization One Laptop per Child, which supplies inexpensive computers to children in developing countries. They did some contract work on solar roadways, which are vehicular travel surfaces that are covered not with asphalt or concrete but with photovoltaic panels. (To test the durability of one prototype, they built a carousel-like contraption in which an automobile wheel on a revolving axel repeatedly rolled over the surface.) They invented a programmable, spoke-mounted LED safety-lighting system for bicycles. They founded the Web site Instructables (www.instructables.com), a compendium of user-described do-it-yourself projects.

Some of these ideas turned out to be dead ends. Electronic rope has yet to find a manufacturer; the hundred-dollar-laptop concept became less intriguing as the prices of commercially manufactured laptops fell; and solar roadways turned out to be what Griffith described to me as "a terrible idea." ("In theory, solar roadways look great—let's just cover all the roads with solar cells—but it's a very energy-intensive way to build a road, and you're unlikely to get that energy back.") Other ideas—the bicycle lighting company, now called MonkeyLectric, and Instructables—evolved into independent enterprises.

The most potentially significant Squid Labs project was spun off as another independent enterprise, a wind-energy company called Makani Power. Griffith, who was the company's original president, is no longer involved on a daily basis, but Makani's general aims are closely aligned with his, and its history elucidates his thinking about both the promise and the limitations of purely technological approaches to dealing with fossil fuels and climate change.

In the mid-1990s, well before he got to MIT, Griffith came across the work of an eccentric early-nineteenth-century English schoolteacher named George Pocock, who invented a kite-powered carriage, which he called a charvolant, and raced it successfully against carriages pulled by horses. Pocock also published a treatise, called *The Æropleustic Art, or Navigation in the Air, by the Use of Kites, or Buoyant Sails*. Griffith's interest in wind was not merely academic. During a trip home to Australia, he saw some kitesurfers, and when he returned to Cambridge he and several friends decided to build their own boards and kites in various MIT labs. They developed their surfing technique by trial and error, and they didn't stop for winter, and Griffith eventually taught classes in kite design and construction, as well as becoming familiar with Boston-area emergency rooms. (He has broken a couple of dozen bones, some of them more than once, and he has scars all over his body, including his face.) You can see a video of one of his more reckless wind-related adventures by searching for "The Stoopid Thing" on YouTube. In the video, the person hanging from the kite, nearly a hundred feet above the water, is Eric Wilhelm, an MIT friend and the cofounder of Instructables; the person on the wakeboard at the other end of the rope is Tim Anderson, another friend, who invented a starch-based 3-D printing technology that made him a multimillionaire; and the person operating the video camera is Griffith, whose turn on the kite immediately preceded Wilhelm's and wasn't recorded. And if you search YouTube for "kite ice butt boarding" you can watch a video of the same three men testing a wind-powered ice sled (which they made in an MIT lab with the help of a tool called a water-jet cutter) and ski-boot ice skates fabricated from old garage-door springs. Both videos bear similarities to the color plates in *The Æropleustic Art.*

"Wind is the second-largest renewable-energy resource, after solar," Griffith told me. "But the strongest, steadiest winds are at altitudes that are higher than you can reach with a conventional turbine. So the question is: how do you get access to that energy?" The world's supply of wind is huge, but it's diffuse: you need big sails to catch enough of it to do useful amounts of work. But structural and economic constraints limit the size of conventional turbine installations, which seldom have towers taller than about three hundred feet. To go higher, Griffith and his friends decided, they needed to get rid of the towers. That was the idea behind Makani. (The word is Hawaiian for "wind.")

Very similar ideas were outlined in the 1970s by Miles L. Loyd, who was a nuclear scientist at Lawrence Livermore National Laboratory. (He's now retired.) During the American oil crisis in the early 1970s, Loyd began thinking about energy alternatives to fossil fuels, and he patented a number of methods by which kites of various kinds might be used to generate electricity. Griffith told me, "He did an astoundingly good job, and nearly everything he wrote, all the physical generalities, he got right." In 1980 Loyd published a technical paper called "Crosswind Kite Power," in which he demonstrated the feasibility of efficiently generating electricity with turbines mounted on unpowered aircraft flying at angles to the wind—a paper that Griffith cited to me as an inspiration and guide. Griffith said, "The system we're building now is not terribly different, in any substantive form, from what Loyd proposed thirty years ago."

30

Harnessing Wind Without Windmills?

Makani's headquarters are on the Oakland side of San Francisco Bay, at what used to be the U.S. Naval Air Station in Alameda. The base closed in 1997 and was declared a Superfund site two years later. Today, it consists of unused runways, crumbling streets, empty parking lots, and postapocalyptic-looking semiabandoned buildings. (When the producers of the television program *MythBusters* need to blow something up, they sometimes do it at the base.) Makani occupies the old air-traffic-control building and tower, at the southeast corner of the runway complex, as well as an adjacent building. When I visited, in 2010, I parked near a rusting gun battery, which was aimed approximately toward Palo Alto.

Research at Makani is directed by Corwin Hardham, a Squid Labs alumnus, who met me on the front steps and led me into a large room. Hardham earned a PhD at Stanford while developing vibration-isolation control systems and actuators for the Laser Interferometer Gravitational-Wave Observatory in Livingston, Louisiana. We stepped over wings, fuselages, rotors, turbines, circuit boards, and other components and looked on as a half dozen young men and women tinkered with prototypes at various stages of assembly or disassembly. Makani's staff, like that of every other venture with which Griffith has been associated, defies common stereotypes about the physical fitness, body mass index, and solar exposure of high-level nerds: almost all the company's employees are accomplished athletes. Hardham, before earning his PhD, considered becoming either a professional windsurfer or a ballet dancer, two other careers for which he was qualified.

He led me through a computer room, where one of Makani's employees was working on a simulation program, and into the building next door, which the company uses as a workshop. He said, "We started with an architecture that is commonly referred to as a winch: you fly a kite, and as the kite flies away from you, in circles, it pulls string off a drum, and the drum is attached to a generator, and, as the drum spins, the generator produces power. Most groups that experiment with what we're doing start in that place. But after we had spent a long time looking at the economics of that, and also at certain aspects of the control of the overall system, we chose to go with a different architecture." Hardham showed me what looked like a large model airplane, with an eighteen-foot wingspan, and said, "We mount the turbines on a rigid wing and fly the wing around in circles and transmit electricity back to the ground through the tether that connects the wing to the ground." Attached to each wing was an electricity-generating turbine about half the size of a coffee can, and each turbine had a propeller-like rotor just over a foot long. On the prototype we were looking at, Hardham said, the two turbines combined, at peak production, produced approximately twenty kilowatts, or enough to power about ten modest American houses. A utility-scale version would be larger—with a wingspan of roughly a hundred feet, or half that of a Boeing 777—and would have a peak generation rate of a megawatt, or enough to power five hundred houses.

We returned to the main building, and Hardham

showed me a video of a recent test flight. In it, a prototype was attached to a cable and flying very rapidly in a circle at thirty degrees to the ground. By flying across the wind, the prototype was able to move significantly faster than the wind's speed—in fact, five or six times as fast, Hardham said—thereby increasing its generating power. Griffith told me that a full-scale Makani-type installation would look "like a wind farm," but higher in the sky, and that, like other wind farms, it would most likely be situated at some distance from densely populated areas, perhaps above agricultural land or the ocean. "You'll just have a bunch of very large kites, flying in circles all day, two thousand feet above the ground," he said. A small kite can sweep the same area as a large turbine, and it doesn't need a tall, heavily reinforced steel tower to hold it up. A standard 1.5-megawatt General Electric wind turbine weighs more than a hundred tons, including the concrete base that anchors it to the ground; a Makani kite with the same output rating, Griffith told me, would weigh three or four tons.

Hardham said, "One way to imagine how all this works is to picture the blades of a conventional wind turbine going around in circles. Now imagine erasing all of that turbine except for the tip of one of the blades. Now imagine tying a string to that tip. What that gives you is a kite flying in circles—and that's essentially what we do. Because we've eliminated the tower and the blades, we're no longer constrained to sweeping a small, tight-diameter circle or to flying close to the ground. We can move our circle up into the sky, where the wind is better, and we can enlarge it so that the swept area is huge."

Griffith said, "The intuitive argument is that it's going to be a lot cheaper to build a four-ton machine than a hundred-ton machine if you're using roughly the same materials—which you are. That lowers the capital cost of your machine and gives you a whole bunch of other advantages." Among the advantages is that the carbon footprint and environmental impact of making the machine itself go down: an airborne turbine contains far less steel, aluminum, plastic, and concrete, as well as smaller quantities of other materials, including the "rare earth" element neodymium, which is used in turbine magnets and is mined mainly in China.

Another advantage has to do with "capacity factor," which is the actual average output of a machine expressed as a percentage of its theoretical maximum output. Wind doesn't blow continuously or at a consistent speed; one consequence is that a standard tower-mounted turbine rated 1.5 megawatts actually generates much less than 1.5 megawatts most of the time. "The capacity factor of a standard turbine is something like 25 or 33 percent," Griffith said. "But as you go higher in the sky, your capacity factor goes up, because the winds are steadier and stronger and your dynamic range is greater. In fact, with Makani technology the capacity factor is more like 60 percent. So what all that means is that the machine is cheaper to make, megawatt for megawatt, and you get that megawatt more often." According to the British physicist David J. C. MacKay, standard tower-based wind generation has a power density of roughly two watts per square meter over land, and three watts per square meter over water. Griffith has estimated that airborne wind turbines have a power density of roughly six watts per square meter—a potentially significant difference, given that even conservative estimates of the world's renewable-energy requirements imply immense installations.

Makani Power began on a boat. In 2006, Saul Griffith and Don Montague—a professional windsurfer and kitesurfer and the holder of a number of patents related to kites, sails, and computer-aided design—invited Sergey Brin and Larry Page, the founders of Google, to join them for a ride on the French trimaran *Geronimo*, which had just set a speed record in the Pacific. That excursion was followed by others, on wind-powered vessels that Griffith and Montague had designed and built, and those outings eventually led to a $10 million investment by Google's philanthropic arm, Google.org, in 2006 and an additional $5 million in 2008, making it Makani's primary financial backer.

Google is interested in energy mainly because the company's server farms, along with the rest of the Internet, use a huge and rapidly growing amount of electricity. Google has revealed very little about its energy consumption, but the total is known to be immense. Searching, accessing, and storing an ever-increasing volume of Web pages, YouTube videos, family snapshots, television programs, e-mails, old books, cloud applications, Tweets, pornography, and everything else that can be found online requires energy, and most of that energy is currently generated by burning fossil fuels. The Internet's energy and carbon footprints almost certainly now exceed those of air travel, Griffith told me, perhaps by as much as a factor of two, and they are growing faster than those of almost all other human activities. In February 2010, the federal government decided to allow a Google subsidiary to participate directly in energy markets, on an equal footing with utilities. In 2011, Google and various investment partners applied for approval to construct two long-distance power transmission lines on the east coast, in a plan to provide wind-generated electricity to urban markets.

Google's investment in Makani is part of an initiative, announced in 2007, called Renewable Energy Cheaper Than Coal, or RE<C. The price of coal is environmentally significant because coal, historically, has been the cheapest and most abundant of the fossil fuels (although in the United States recently natural gas has competed with it as a bargain), and no genuine transition to renewable energy will be possible without doing something to eliminate that market advantage, either by increasing coal's cost—through the imposition of carbon taxes, the elimination of coal-industry subsidies, the recognition of so-called externalities (like mining deaths and coal-related health costs), or other government actions—or by developing renewable-energy sources that can compete with fossil fuels on price alone. The first approach appears to be a political

impossibility; the second, in the absence of the first, may be a technological one.

Griffith and Hardham believe that Makani's innovation, by generating power more efficiently than conventional turbines, could narrow the difference. Maybe they're right; maybe they're screwy. But there's a bigger, more difficult question: if their kite design or any other innovative, unconventional energy concept has genuine global potential, how will we ever find out? For all our talk about the importance of innovation, there is no clear path from possibly cool idea to global implementation.

31

The Discouraging Economics of Innovation

In 2010, the U.S. Department of Energy's Advanced Research Projects Agency-Energy (ARPA-E)—which was created by Congress to invest in "creative, out-of-the-box, transformational" energy research—granted Makani an additional $3 million, which the company is using to fund further proof-of-concept testing, including work on taking off and landing. Yet moving Makani or any other innovative energy company beyond the promising-prototype stage represents a tremendous challenge, because the required amounts of money are huge. In early 2010, Griffith told me, "We are more advanced than the Wright Brothers were—we know we can fly, we know everything can work—but you're not really in the game until you're at the DC-9 stage. We've got past Kitty Hawk, but the technology is still at small scale, and it still doesn't work every single time, and it's expensive. We know we can make it fly, generating power, for twenty-four hours in a row, but if you're a utility buying electricity you don't want something that has only ever worked for twenty-four hours, and only at a small scale. We need to show that this machine is going to work at enormous scale, and that it will stay in the air and conform to specification for more like twenty thousand hours."

Doing that would require a very large financial commitment—far larger than the company has been able to attract so far. And that's true of any potentially significant energy technology, including the brilliant-sounding ideas that you read about in the newspaper on a daily basis—all the schemes involving algae or switch-grass or restaurant waste or geothermal wells or magnets or quantum solar panels. Griffith continued, "Even if I came to you tomorrow with the perfect energy idea, the reality is that to go from that idea to doing utility-scale power generation would be a minimum cost of entry of a hundred million dollars, and a minimum lead time of five to ten years."

And even then, in return for your investment you'd

have only a single generating plant, which might or might not live up to its promise. Risk on that scale puts any such project far beyond the reach of a rational venture capitalist, or even of a well-capitalized and highly motivated company like Google. Joby Energy, a small private company in Santa Cruz, California, is pursuing airborne-turbine technology similar to Makani's. It has been financed primarily by its founder, JoeBen Bevirt, who has earned millions from earlier ventures of his, including the company that sells the GorillaPod camera tripod, which he invented in 2005. (Bevirt and Griffith are friends.) But that bankroll, though considerable, will not be sufficient to carry Joby's wind technology to utility-scale deployment. Joby faces the same problem Makani does: who is going to finance the leap from possible promise to national grid? Makani Power, Joby Energy, and innumerable other innovative small companies, inventors, and entrepreneurs are developing tantalizing and potentially valuable approaches to energy production, but their next steps are by no means clear—partly because the global recession, reduced electricity-generation costs, and drifting public attention have weakened any sense of urgency about renewables, and partly because no obvious financial mechanism exists to take such companies from prototype to industrial implementation. In 2011, Michael Levi, in his blog on the Web site of the Council on Foreign Relations, wrote, "Venture capital has become pretty much synonymous with financial support for cutting-edge technology. But the established VC model—relatively small scale of capital, 3–5 year investments—is a pretty poor match for much of the energy field, in which early stage projects will take much longer to mature and will cost a lot more money than the VC norm. The sharper analysts who understand this tend to conclude that government will need to step in to make sure technological progress happens. Perhaps. But the history of government support for innovation aimed at commercial application isn't all that encouraging. We'd be a lot better off if someone invented a financial structure that allowed investors to collectively take longer-term, higher-capital risks." Levi acknowledged that his idea was possibly "a pipe dream," but he also observed that "venture capital basically didn't exist seventy years ago, and didn't really take off until the 1980s." Still, the difference in scale is huge, and it's not clear that any nongovernment funding source is big enough to bridge it.

* * *

News reports about energy research can create the impression that our power problems are just about licked, and on numerous fronts. But the reality is different. The speed with which software-based activities and Web innovations catch on—Amazon, eBay, Facebook, iPhone apps, Groupon—has encouraged a public perception that transformative technological change takes place almost instantaneously. One awesome idea while you're a sophomore at Harvard, and—boom!—you're a billionaire and your picture's on the cover of *Time*. But hardware is harder and slower than software. It also has far bigger capital needs, requires longer attention spans, entails bigger financial risks, and, when successful, generates smaller, less immediate returns.

This is usually true even for relatively small-scale hardware innovations. Griffith was initially drawn to MIT by his interest in reducing the amount of paper in the global waste stream. (Paper constitutes between a third and a half of the content of landfills, worldwide.) A science-fiction book called *The Diamond Age*, by Neal Stephenson published in 1995, got him thinking about low-energy digital alternatives to printing on paper—in the novel, people do most of their reading on page-thin "mediatrons"—and an article in *Wired* led him to MIT's Media Lab, where research was then being done on electronic ink, the technology now used by the Amazon Kindle and similar digital readers. By the time Griffith got to Cambridge, the fundamental work on electronic ink had ended, but he contributed to the final stages, and Joseph Jacobson, one of the technology's inventors and a cofounder of the company E Ink, became his thesis adviser.

That was a dozen years ago—an eon by software standards. (During the same period, Napster, Friendster, GeoCities, and Ask Jeeves have come and, for the most part, gone.) Yet the amount of paper in the waste stream remains huge, and E Ink's share of the global reading market, though it's grown impressively, is still extremely small. Furthermore, Apple and a rapidly lengthening list of other device manufacturers are now heavily promoting tablet computers—which, because their electronics are more complex and their displays are back-lit, use more energy and other resources than any electronic-ink device—and are clearly hoping that their products will turn the Kindle and similar dedicated readers into landfill. And lots and lots of brilliant people are working very hard, right now, to turn those products into landfill, too. I attended the 2011 Consumer Electronics Show in Las Vegas and was staggered by the acres and acres and acres of clever high-tech gizmos—few of which, I knew, would still be for sale in a year or two. Much of what we think of as innovation is more like technological churn. "Built-in obsolescence" doesn't come close to describing it, since most of the devices we covet today are technologically extinct even before we've decided we can't live without them.

32

Getting from Lab to Grid

ARPA-E—the federal agency whose mission is to finance high-risk energy research—was founded in 2007 and first funded in 2009, when it was allocated $400 million as part of the George W. Bush administration's economic stimulus bill. That's real money, but the agency's portfolio is extremely broad, and most of its awards have been small. (Grants have generally fallen between $500,000 and $4 million and have usually been paid out over more than a year.) In addition, the money is spread across many fields and concepts, and most awards have gone into research areas, such as energy storage and improved air-conditioning, that were already well funded by private industry or by other parts of the federal government. ARPA-E has received further funding since then, but these basic issues remain.

A little over fifteen years ago, I wrote a magazine article about a man who had founded a company to promote an unusual technique for hitting golf balls—a technique that he believed to be simpler and more effective than standard methods—and to sell golf clubs that he had designed to complement his unusual swing. A couple of years later, his main investor forced him out of the company, which was struggling. I offered condolences, but he said not to worry. All the profit in a company like that, he said, is in founding it and finding investors, not in running it. And he moved on to something else.

Much the same can be true of "creative, out-of-the-box, transformational" energy research, because, for those who engage in it, most of the potential returns take the form not of long-term operating profits but of grants, subsidies, research stipends, small-scale speculative investments, and tax breaks, or of profits from selling interim products of questionable environmental value, such as solar panels for suburban houses or algae-based cooking oil. The federal government, green-minded venture capitalists, and global energy companies that are concerned about their public image will never lack for interesting-sounding places to throw money—better battery chargers for cars, improved power supplies for tablet computers, smartphone apps for monitoring household electricity use—but small investments like those are too diffuse to have a noticeable impact on the world's real problems, which are measured in hundreds of billions of dollars and on which progress is constrained less by a shortage of cleverness than by a lack of public will. Meanwhile, the one truly concerted American investment in renewables—corn-based ethanol—has, by all reliable accounts, actually worsened the nation's energy and climate predicament, while pushing up food prices both here and abroad.

The human race, Saul Griffith has estimated, currently consumes energy at an average rate of approximately 16 trillion watts, or sixteen terawatts—the equivalent of 160 billion hundred-watt lightbulbs burning all the time. Capping atmospheric greenhouse gas at 450 parts per million—a level that's 15 percent higher than today's and that climatologists hope may be consistent with a global temperature increase of only about two degrees Celsius—would necessitate freezing global energy consumption at the current level (despite a projected worldwide population increase, by midcentury, of roughly 2 billion people, or more than six times the current population of the United States)* and replacing all but three of those sixteen terawatts with energy generated from a combination of the most promising renewable and non-carbon-based sources: photovoltaics, solar thermal, wind, biofuels, geothermal, and nuclear fission. And doing that, Griffith said, would require building the equivalent of all the following: a hundred square meters of new solar cells, fifty square meters of new solar-thermal reflectors, and one Olympic swimming pool's volume of genetically engineered algae (for biofuels) every second for the next twenty-five years; one three-hundred-foot-diameter wind turbine every five minutes; one hundred-megawatt geothermal-powered steam turbine every eight hours; and one three-gigawatt nuclear power plant every week. Such a construction program, he told me, is at least theoretically achievable, but the practicalities are daunting. There is no remotely

* More realistic forecasts assume that global energy use will actually double.

plausible plan for a low-carbon world, for example, that doesn't include huge increases, worldwide, in the number of nuclear power plants. Yet the design, approval, and construction of nuclear plants can take decades, and proposed facilities in this country and elsewhere face intense public opposition, especially following the 2011 disaster in Fukushima, Japan. In the United States, the number of functioning nuclear plants, rather than increasing at anything like the rate of one per week, is actually falling: no construction of a new U.S. plant has begun since 1977, and no new U.S. plant has gone online since 1996, and utilities and regulators, thus far, have shown more interest in decommissioning existing facilities than in building new ones. In Germany in 2011, Chancellor Angela Merkel announced that by 2022 the country would shut down all seventeen of its nuclear reactors.

"Right now, everyone sees climate change as a problem in the domain of scientists and engineers," Griffith told me. "But it's not enough to say that we need some nerds to invent a new energy source and some other nerds to figure out a carbon-sequestration technology—and you should be skeptical about either of those things actually happening. There are a lot of ideas out there, but nothing nearly as radical as the green-tech hype.

We've been working on energy, as a society, for a few thousand years, and especially for the last two hundred years, so we've already turned over most of the stones."

In this context, a commitment to "transformational" innovation can seem like stalling. If we're really serious about making a transition to low-carbon and renewable energy, we should begin the transition, not halfheartedly window-shop for miracles. New utility-scale wind, solar, and nuclear installations employing current technology probably have a useful life of a couple of decades. Rapidly building a significant number of them would, in addition to everything else, create a moving twenty-year window in which to decide which "creative, out-of-the-box, transformational" ideas to replace or supplement them with—or to say the hell with it and go on as before. We accomplish nothing by, in effect, waiting for Caltech and MIT to hand us the perfect solution. The world's primary climate-and-energy strategies consist of arguing with one another and spending money in ways unlikely to accomplish anything significant or lasting.

33

Retrograde Innovation

Saul Griffith's "do tank," Squid Labs, lasted just three years, but its demise didn't destroy Griffith's enthusiasm for private laboratories. "It's not an easy thing to do and survive at," he told me, "and, over time, the temptation is to become a very vanilla consulting company. But in the heyday of American invention there were a lot of private labs, like Edison's and Tesla's, back before universities became the only places to do research." In 2009, he founded a scaled-down successor, called Other Lab, with two partners: James McBride, who studied physics and quantum computing at MIT and later created financial trading systems in New York, and Jonathan Bachrach, who is a scientist, a software designer, and an electronic artist. Bachrach and Griffith met at MIT, in a class on amorphous computing, which concerns intelligent systems whose processing power is not contained within a central component but is dispersed among a large and, often, fluctuating number of irregularly interconnected elements, none of them particularly intelligent. (Griffith defined it for me as "how you make a fish tank think.")

Other Lab occupies a storefront in a low, modern building in an industrial neighborhood of San Francisco near Potrero Point. When I visited, in late 2009, Bachrach let me in, then propped the door open with a gallon jug of wood glue. An old wooden kayak was hanging from the ceiling, and bicycles, bicycle parts, and bicycle tools were everywhere. Griffith seems to operate on the principle that excessive orderliness is inefficient and that neatly putting things away is more time-consuming, in the long run, than searching through piles. Among his few concessions to conventional housekeeping are a half dozen salvaged library card-catalog cabinets, the drawers of which he has repurposed as (nonalphabetized) storage units for thousands of small parts: strap clips, eyelets, resistors, microcoils, standoffs, shackles, hose clamps, bearings, springs, washers, cleats, skate hardware. Parked near the door was the prototype for an electricity-assisted tricycle, an ongoing Other Lab project. It had a yellow barrel-like enclosure mounted in front for hauling cargo. That morning, the cargo had consisted of Huxley Griffith, Saul and Arwen's infant son. During Griffith's ride to work, rain had caused a short circuit in the wiring near the trike's battery, and a cloud of gray smoke had emerged from under Huxley's seat. "There was this hissing sound, and I had to pull him out and try to stamp out the fire," Griffith said. Huxley had reacted placidly to the crisis, as though, at eight months, he was already accustomed to life as the child of an inventor. The trike is typical of a number of Griffith's recent inventions, in that he designed it to address a perceived environmental issue in his own life—dependence on automobiles—while also hoping eventually to find a market for it. "The hypothesis is that, by being completely selfish and solving all my own energy problems, I will find some general solutions that other people will like, too," he said.

The trike also presents an energy problem of its own. Its battery—which can provide about a kilowatt-hour of power on one charge, or enough to give the trike a range of about fifty miles, assuming the rider pedals half the time—costs a thousand dollars, or as much as the rest of the components combined. "So, effectively, energy storage doubles the cost of the bike," Griffith said. "It's the entire problem of electric vehicles and hybrids." No battery comes anywhere close to holding energy as efficiently as the gas tank of an ordinary car. ("Unfortunately," Griffith told me, "the difference between the world's best battery and gasoline, in terms of energy storage per kilogram, is not a factor of ten; it's more like a factor of hundreds or thousands.") This is a critical issue, because the most abundant renewable-energy resources are intermittent—the sun sets; the wind stops blowing—and so rely on some form of stockpiling. "The French, during the night, store energy from their nuclear power plants by pumping water uphill," Griffith said, "and then that water is used to generate power hydroelectrically the next day, when demand is high again. So that's like a huge battery. But there's an issue of what's known as round-trip efficiency. Most energy-storage systems have round-trip efficiency of about 80 percent, meaning that you lose 20 percent as you store the energy and then retrieve it. So in order for wind and solar to work we also need to do storage at enormous scale."

He rummaged around on a large table in one of

Other Lab's two workrooms and showed me a heavy black iron device the size and approximate shape of a small loaf of bread. It had a hand crank, which he turned. "This is a power supply for an old telephone, circa 1920," he said. "It's almost a century old, but it still works. If you put your tongue on there, it will throw you across the room. And you could keep it working, conceivably, for another two hundred years." Cell phones can't do that, because they depend on energy-storage components called electrolytic capacitors, which deteriorate over time, and because their batteries become useless after a finite number of charging cycles. "Technology optimists would say, okay, we'll invent a better battery and better electrolytics," Griffith said. "But the possibilities for big improvements probably aren't that great, and, besides, there's a better way to solve the problem, which is to fundamentally rethink the design of the object, along with the sociology and the behavior around it. Example? Hand-crank your cell phone."

Other Lab's coffeemaker is a hand-operated La Pavoni espresso machine, on which Griffith made coffee for both of us. The design is a hundred years old, and the machine itself, with occasional repairs, could easily last a century, he said, since it has no complex or nonmechanical parts. "But a modern espresso machine that you buy at Crate & Barrel has a liquid crystal display and a little computer and microcontroller and a whole bunch of electrolytic capacitors, and it probably won't last ten years." And when a coffeemaker like that goes bad it can't easily be fixed—a weakness that affects almost all the fancy devices we now depend on and that has contributed to the near extinction, in the United States, of an entire species of commercial enterprise: the general-purpose repair shop. (Most Americans under a certain age don't even realize that there might be something to do with a malfunctioning toaster oven or television set—or pair of shoes with worn soles—other than to throw it away and order a new one.) Even when modern appliances can be repaired, the cost is often daunting. When the motherboards on the side-by-side ovens in my wife's and my kitchen self-destructed simultaneously—a victim of heat from the ovens themselves—replacing the fried electronics cost hundreds of dollars, even though the parts themselves were covered under warranty. And it's now virtually impossible to buy large appliances that don't contain similarly fragile high-tech components.

Such vulnerabilities have implications for the so-called smart grid, a concept that plays a large role in almost all plans for a postcarbon world. The basic idea is that significant energy savings could be achieved if information flowed readily in both directions between energy users and energy suppliers. Smart appliances, for example, might switch themselves on only during off-peak power-consumption periods, enabling utilities to level out peaks and valleys in demand. Or they might cycle on when the sun was shining on solar arrays, and cycle off when wind turbines weren't turning. The presumed benefits sound tantalizing, but they depend not only on nondisintegrating computer chips but also on an ability to make sense of digital information generated by millions of dissimilar machines. My wife and I recently bought a new high-definition television set and a Blu-ray disc player and received a new digital video recorder from our cable company. Successfully operating all three devices seldom requires fewer than two remote controls, and sometimes all three, even though one of the three is supposedly "universal." Solving the remote-control problem—which has existed for decades—ought to be child's play in comparison with the far greater challenge of making our entire power system function intelligently, since remote controls involve simpler technology and many fewer elements. But the problem is still with us.

Sometimes it's hard to believe that the smart grid isn't at least partly a marketing scheme devised by manufacturers. At the 2011 Consumer Electronics Show, I attended a seminar on networked appliances and realized—as the panel's members described the wonders awaiting homeowners savvy enough to upgrade to smarter refrigerators, dishwashers, and clothes dryers—that I'd heard it all before, and not just once. On my smartphone I searched the *New York Times* online and found an article that closely tracked the discussion in the seminar room. "Think . . . about calling up your appliances—the refrigerator, the hot tub, the alarms—from the car phone as you commute home from work," the article said. "The refrigerator defrosts a pie and tells the oven to start the roast; the range signals the microwave oven to heat the souffle, and 102-degree water fills the bathtub. As you drive under the automatic garage door, the lights switch on, the heat fires up, the security system turns off and Andre Watts plays from a compact disk." That article, by Joseph Giovannini, was published in 1988. "Car phone," "Andre Watts," and "compact disk" are giveaway anachronisms, but the rest of the text I've quoted could serve as a close summary of the discussion I attended at CES. Because this was 2011, the main benefits for consumers were said to be energy savings, rather than Jetson-style convenience. But the pitch was otherwise the same.

All this is less of an issue than it might be, since we tend to replace still-functioning devices with improved versions long before their electronic innards have gone kaput. Nobody who upgraded to an iPhone 4 did so because their iPhone 3 had stopped working. And the much-derided decision by Apple to make the iPhone's battery nonreplaceable has been borne out by the eagerness of customers to unload superseded models while their batteries were still functioning at close to their original capacity.

Even gadget-loving consumers are often made uneasy by the speed with which we acquire and abandon complicated, expensive possessions. (David Pogue, the *New York Times*'s technology columnist, observed in 2011 that hardly any of the devices he'd reviewed during the previous decade were still on the market.) And when we feel uneasy in that way we usually blame manufacturers, for introducing irresistible new models before we've even broken in the old ones. But, of course, the real problem is us. And we're not a small problem, either, since in one way or another we all depend economically on the continued fickleness of everyone else. It's one thing to say we feel overwhelmed by our junk; it would be quite another to demand societal changes whose direct results would include a steep decline in economic activity, leading to reductions in our own income, comfort, and convenience.

If we ever do find ourselves in the mood for such radical changes, there actually exists, close at hand, a worked-out model for how to manage consumer technology in a less environmentally devastating way: the old AT&T, a monopoly that endured more or less intact into the 1970s. The "Bell System," as it was known, had no competitors, so there was no headlong rush to build and abandon costly, tantalizing, resource-intensive infrastructure or to implement services that would become obsolete in the next wave of innovation. And AT&T itself owned all the phones, which it rented to customers at a hefty markup—giving it an incentive to keep handsets simple, boring, and extremely durable, like Griffith's old telephone magneto. Of course, because AT&T had no competition, telephones were far less interesting and useful in those days, and phone calls were often shockingly expensive. (The concept of the "long-distance call" has virtually disappeared from American life, as have dialing-related finger injuries.) But, for anyone who wrestles with the environmental consequences of accelerating replacement cycles, the Bell System is worth pondering. If you couldn't buy electronic gadgets, but had to rent them from Apple, Apple would make sure that your (hand-cranked) iPhone lasted for decades, and it would be a very long time before you were offered an opportunity to abandon your iPad for an iPad 2. That's a grim thought to anyone who has grown accustomed to rapid cascades of high-tech upgrades, but it's a comparatively green one. The question is whether we could ever feel sufficiently threatened by our own appetites to voluntarily embrace anything so boring. How appealing would "green" seem if it meant *less* innovation and *fewer* cool gadgets—not more?

34

The Conundrum

In the United States, being an ideological purist on environmental matters was easy during the first eight years of the twenty-first century. The George W. Bush administration was so hostile to almost anything resembling a green initiative that, for as long as Bush was in office, few big ideas had to be tested against reality—political, economic, scientific, or otherwise. The election of Barack Obama, in 2008, made many people who worry about environmental issues believe that a major obstacle to serious American action (and therefore to global action on a meaningful scale) had been removed.

But talking about change, as always, turned out to be easier than effecting it or even agreeing about what to do. During Obama's first two years in the White House, some of the fiercest conflicts were not between climatologists and "deniers" but between environmentalists and environmentalists—and those conflicts have, if anything, become more intense. The problem is seldom that one group has all the facts and everyone else is deluded; as frequently happens with complex issues, the biggest impediments to effective action have been truths, not falsehoods, and the fiercest arguments have often been between ostensible allies. Hydroelectric power is substantially emission free, but dams destroy ecosystems and human communities. Uranium-235 has a small carbon footprint, but what about accidents, earthquakes, terrorists, and nuclear waste? American gasoline is one of the cheapest manufactured liquids in the world, even today, but taxing it more heavily would increase unemployment and deepen the recession. Compact fluorescent lamps use less electricity than incandescents, but contain mercury. (The U.S. Environmental Protection Agency says that, if the powder from inside a broken compact fluorescent gets on your clothing, you should not wash the clothing but instead seal it in a plastic bag and throw it away.) Photovoltaic panels and solar-thermal concentrators have potential as electricity sources, but building utility-scale installations would ravage the deserts that are the ideal places to put them. Some of the most vocal opponents of the Cape Wind project (a proposed wind-turbine installation in Nantucket Sound that received a federal lease in 2010, after a decade-long fight, and that, if it's constructed, will be the first American offshore wind facility) have been environmentalists.

Tea Party–influenced Republican victories in the midterm elections of 2010 were a setback for Obama and for those who, as recently as a year or two before, had hoped for sweeping American initiatives on energy and climate. But to some of those same disappointed people the elections were also comforting, because they returned many vexing environmental issues entirely to the realm of the abstract. Debating the potential benefits and hazards of a global program to rapidly build vast numbers of thorium-fueled nuclear reactors (for example) is harder and more frustrating than voicing outrage over the latest pronouncement of Sarah Palin or John Boehner, who, in 2010, suggested that he thinks the scientific brief against carbon dioxide is that it causes cancer.

A few determined ignoramuses (abetted by lazy reporters, broadcasters, and bloggers) can cause plenty of trouble; witness the public-health crisis exacerbated by the former *Playboy* centerfold Jenny McCarthy and her campaign against childhood vaccinations. But if environmental troglodytes were the only obstacles to global action on energy and climate our challenge would be less daunting than it is. The real problem isn't them; it's everyone—especially those of us who, however enlightened we may feel, are quite comfortable consuming a grotesquely disproportionate share of the world's resources. Just how willing are we, actually, to demand and support policies that would require more from us than product substitution? And, even if we believe we are willing, how could we prevent the biggest burdens and sacrifices from falling on those who live in misery even now?

Our climate-and-energy dilemma is really a world-sized version of the tragedy of the commons. It would be in the collective interest of all parties to manage the earth's resources for maximum longevity—that is, to manage them for sustainability, or for what William Stanley Jevons called mediocrity. Yet as consuming individuals we have a conflicting but irresistible short-term interest in grabbing as much as we can, even if by doing so we bring eventual harm to ourselves and our descendants—

since we personally enjoy all the benefits of our own consumption, yet share the consequences with everyone else, and primarily with people not yet born. That's a conflict that "the market" can't resolve. Supply-and-demand is a powerful force, and for the world's fortunate minority it has generated extraordinary comforts. But free markets aren't always benign: sometimes the invisible hand goes for the throat. Ruined aquifers are a market response to diminished resources; so are famine, terrorism, and civil war.

Still, economics is the only lever we possess that's long enough to move the world. Do we have the courage to use it as an instrument of sacrifice? We know how to make people consume less: charge them more. We also know how to reverse population growth, how to produce energy without fossil fuel, and how to restrain wasteful consumption. What we don't know is how to make ourselves do those things on a global or even a national scale, with all parties in approximate agreement; we certainly don't know how to make it all work equitably, across the full range of global affluence levels, and in perpetuity. Nor do we know how to anticipate and prevent harm we don't intend. America's main renewable-energy initiative, its misconceived investment in biofuels, has increased global hunger, by drawing cropland out of food production; the European Union's cap-and-trade program—according to a study published in *Energy Policy* in 2011—has perversely encouraged the consumption of coal.

These are not problems that scientific innovation can solve, because they have nothing to do with technology; they have to do with us. And whether we truly have the ability to deal with them is something we can't know until we've actually tried. The likelihood of our doing that is probably low, but a few points seem almost axiomatic:

- For the world's most affluent people—and especially for Americans, Canadians, and Australians—energy is too cheap. Making it more expensive for the wealthy, through some comprehensive system of equitable taxation, would push down consumption and make investment in utility-scale renewables economically rational. The blow to individuals and businesses could be softened by finding some way to return tax revenues, as rebates, at rates high enough to prevent suffering but low enough to preserve an incentive for reduced consumption. (If you paid ten or fifteen dollars for a gallon of gasoline but received four or six of those dollars back as your share of the national rebate, you could still get to work but you'd have a clear incentive to carpool or take the bus; if you lived in a dense city and didn't own a car, your share of the gasoline rebate would help you pay the rent on your extremely efficient but shockingly expensive apartment.) Or energy could be priced on a sliding scale, so that it became more expensive to individual consumers as they used more of it—a form of rationing. There are many semi-plausible seeming ideas for doing something like this.

- The real difficulty will be getting a critical mass of people to agree to any scheme and then—gulp—finding a way to extend it equitably across national boundaries, and across socioeconomic differences that dwarf those within the United States.

- All energy supplied to the world's affluent minority—not just fossil fuels, and not just energy used to do bad things—needs to be more expensive. The solar-generated electricity I use to recharge my electric car is every bit as much of an environmental problem as the conventionally generated electricity I use to run the half-empty refrigerator in my basement. Meanwhile, energy for the world's impoverished majority needs to be cheaper and more readily available. How?

- One way to tax energy in wealthy countries might be to establish a schedule of (steadily increasing) prices for energy from all sources, extending well into the future, and then impose taxes to bridge the differences, daily, between individual market prices and those targets. One advantage of such a system is that it would facilitate long-term planning for energy costs, since future prices could be known years in advance. Manufacturers now complain, with justification, that consumers lose all interest in conservation the moment energy prices dip; predictability might encourage rationality. Even so, figuring out how to actually implement anything like this would be an insanity-inducing nightmare.

- Subsidizing energy consumption by the wealthy ("Cash for Clunkers"; tax credits for suburban solar panels) is a doomed environmental strategy.

- Promoting energy efficiency without doing anything to constrain overall energy consumption will not cause overall energy consumption to fall. In fact, if we ever actually do find the will to purposely make energy more expensive for ourselves, we'll have to be equally assiduous about steadily raising those prices, to keep the environmental benefits of reduced consumption from being subverted by our ingenuity at devising new efficiencies. Improving efficiency can enable us to live well on less, but it's self-negating if it merely acts as a consumption amplifier, as it has historically.

▪ One of the least meaningful and most overused words in the English language is "sustainability." For most Americans, it means something like "pretty much the way I live right now, though maybe with a different car." A good test of any activity or product described as sustainable is to multiply it by 300 million (the approximate current population of the United States) and then by 9 or 10 billion (the expected population of the world by midcentury) and see if it still seems green. This is not an easy test to pass.

▪ Permanent, year-over-year economic growth, fueled by steadily increasing consumption of energy and natural resources, is not sustainable in any sense, even though economists tend to treat growth and the environment as nonconflicting categories.

Indeed, they often argue that we can solve our energy, climate, resource, pollution, poverty, and global-equity problems by growing our way out of them. This is the equivalent of believing in perpetual-motion machines and Ponzi schemes.

▪ Income disparity is a global generator of environmental harm. The economist Robert H. Frank, who teaches at Cornell and writes a column for the *New York Times*, has argued persuasively that increases in income inequality in the United States in recent decades have hurt almost everyone, by setting off, even at lower income levels, what he calls "positional arms races"—ultimately disastrous efforts to keep up with the Joneses. In the United States, Frank argues, the increasingly oversized mansions of increasingly overpaid CEOs helped to pull the entire housing market upward behind them, setting off an "expenditure cascade" among middle-class homeowners, many of whom had to stretch to meet their mortgage payments even when times were good. This resulted not only in a huge increase in consumer debt and a sharp decline in economic stability, but also in the doubling of the size of the average American house and, with it, increases in consumption and waste in all categories. Any truly effective environmental strategy will necessarily "soak the rich," both nationally and globally. No humane scheme for truly addressing climate change is compatible with a global socioeconomic hierarchy in which a very small percentage of the earth's people continue to consume a very large percentage of its resources.

- Dense, efficient, intelligently organized cities are the future of the human race, and they provide the only remotely plausible template for permanently housing large populations. But big, dense cities are not all equal. Hong Kong is an environmental paragon; Dubai, which from a distance appears similar, is an environmental disaster.

- Global Environmental Enemy No. 1 is the automobile, no matter what it runs on. The ecological advantage of electric cars—to the extent that there is one—is not that they're ultimately powered by coal, natural gas, or uranium, rather than by oil; it's that their range (so far) is limited, so that they can't easily be driven long distances. What we drive is far less environmentally significant than the rate at which we spin the odometer, because enhancing mobility amplifies all our environmental impacts. Any so-called green scheme that makes you happier to be a driver is both delusional and counterproductive. So is any technological innovation that pushes access to automobiles ever further down the worldwide income ladder.

- Forget about high-speed passenger trains and jets that fly on vegetable oil. The environmental problem with mobility isn't miles per gallon; it's miles. American rail enthusiasts often point to the Chinese as green-travel role models, because they have invested hundreds of billions of dollars in fast-train infrastructure. But China's goal in building its rail network is not to reduce national energy use; its goal is to increase it, by stimulating consumption of all kinds, in all parts of the country. That will be good for China's economy—which is another way of saying that it will enable China to consume more fuels and other resources—but it isn't "sustainable" in any sense.

- We haven't found the ideal ways to generate energy from noncarbon sources, but we know enough to be doing more than we're doing. Too much of our current effort, worldwide, is wasted on disparate, small-scale, miracle-seeking research studies and "demonstration" projects, which provide incomes and interesting activities for university professors, venture capitalists, public-relations firms, lecturers, panel moderators, government agencies, corporations, nonprofit organizations, graduate students, green-minded journalists, sustainability officers, efficiency consultants, LEED professionals, and millions of others—the global eco-industrial complex—but have little chance of accomplishing anything of long-term environmental value. A more useful "demonstration" would be to make a genuine effort of some kind, on a truly large scale, using technology we currently understand—say, a solar and wind installation that supplied 100 percent of the electricity consumption of metropolitan Phoenix. Even most environmentalists would dismiss such a proposal as absurdly ambitious and impractical, despite the fact that it represents a tiny fraction of what we need to do. Yet attempting a truly large project is the only way to honestly assess the actual obstacles to supplanting fossil fuels. Hype, hucksterism, and faulty arithmetic are easy to conceal when the focus is suburban rooftops.

- Creating green jobs is easy; a few tweaks here and there, and almost anyone qualifies. What's hard is eliminating brown jobs—a category for which, again, almost every American qualifies, no matter what they do.

- A seldom-discussed climate issue is the carbon footprint of reducing the world's carbon footprint. Every plan for shrinking production of greenhouse gases (other than directly reducing consumption) involves the construction of vast amounts of new infrastructure for generating and distributing energy, as well as manufacturing huge quantities of "sustainable" products to consume the energy thus produced—and all that new construction and manufacturing necessarily involves the burning of fossil fuels and, therefore, the addition of carbon to the atmosphere. Saul Griffith told me, "There are about 800 million cars in the world today, and that number is projected to go to 2 billion. Now, imagine building 2 billion electric cars, along with five hundred million 'green homes' and the infrastructure to generate ten or twelve terawatts of energy from solar, wind, geothermal, biomass, and nuclear energy sources. The CO_2 emissions from building those things will add twenty to twenty-five parts per million to the atmosphere, or at least a third of the difference between where we are now and 450 parts per million. And that's before you turn even one of the cars on."

▪ Access to freshwater may become an acute global problem before access to energy does. And many low-carbon energy sources, including nuclear and solar, are heavily dependent on large, uninterrupted supplies of water.

▪ True sustainability involves more than technology and economics. In an email in 2011, Herman Daly wrote, "Why one cares about anything is fundamentally a religious question. If the world and its inhabitants are merely ephemeral random improbabilities realized over billions of years of atoms in motion, somehow followed by struggles for reproductive success by randomly emergent organisms, and values are nothing but survival-selected strategies of our selfish genes to reproduce, etc., etc., then why care? If the world is not a Creation but only a temporary 'Randomdom', why not have fun and tear it up?"

▪ Saul Griffith believes that, at the individual level, a meaningful response to environmental challenges consists of choices about what he calls "personal infrastructure": "Which car? Which house? Which job (physical location in relation to the house)? Which large appliances? Which HVAC system? Which diet (vegetarian/omnivore)? To make genuine, long-term, substantive changes to your energy consumption, you have to intervene at this infrastructure level, in decisions that typically arise only every five or ten or twenty years. Everything else is window dressing."

- Nevertheless, individual acts are insufficient—and if they're guided by conventional thinking they're often counterproductive. We learned in grade school that every little bit counts—every act of kindness, every penny saved, every vote cast, every soda can retrieved from the trash. But, in truth, plausible solutions to the world's present difficulties are measured only in billions and trillions. David J. C. MacKay has written, "The mantra '*Little changes can make a big difference*' is bunkum when applied to climate change and power." The only relevant changes, MacKay writes, are truly big ones. Those are the kind even caring individuals tend not to like. The real sacrifice required of each of us is to demand and support change by all.

- When thoughtful Americans brood about climate change, fossil-fuel consumption, and environmental devastation, they often focus their anxieties and scorn on China, which has more than four times the population of the United States. But the carbon footprint of the average Chinese, even now, is just a sixth as large as the carbon footprint of the average American, and China's rate of population growth is only about 20 percent of ours. Both countries are expected to add roughly 100 million new residents between now and approximately midcentury. From an environmental perspective, the world still has more to fear from the Americans. Between now and midcentury the population of the world is expected to rise by somewhere between a third and a half of what it is today—i.e., by about two Chinas, or by six or seven United States of Americas—and global energy demand is expected to double.

- We Americans, cumulatively, now account for more than a quarter of the world's consumption of coal, oil, and natural gas; individually, we consume resources at five times the global rate. No U.S. environmental initiative currently in the works or under serious discussion seems likely to shrink our aggregate impact anytime soon, not least because reductions in the volume of emissions would have to outpace increases in the number of emitters. Yet population control has largely been a forbidden topic in the United States, for both the right and the left—presumably because limiting the number of people, inevitably, involves doing things to limit the number of people, and none of those are popular. China, with its three-decades-old one-child policy, has made a significant reduction in its rate of population growth—but that policy has required a determined central government and a compliant nation, and it's under growing social stress. China's first generation of one-child parents is now reaching retirement age, and, as a result, the country has to figure out how to support all those seniors with a sharply reduced supply of young workers. China's experience is cited far more often as proof that population control is disastrous than as proof that it's achievable. (The brilliant Swedish statitician and demographer Hans Rosling has observed that the only part of China to have actually achieved a birth rate of one child per woman is also the only part of China without a one-child policy: Hong Kong. This suggests that the most effective global population limiter may be affluence—plus urban living and population density.)

- There is a widespread feeling, even in the West, that it would be unfair to suggest that the citizens of less prosperous but rapidly developing countries should forgo conveniences and luxuries that we Americans have come to think of as our birthright. But it's really the residents of those rapidly developing countries who have the most to lose by following the worst elements of America's century-old example. Automobile-based cultures, no matter what they use for motor fuel, are clearly unsustainable. Countries with expanding economies would be better off if they used their new wealth to erect ways of life which they might hope to sustain beyond the automobile's inescapable endpoint, rather than recklessly investing in a future without a future. Not jumping off a cliff is easier than turning around in midfall.

- There's a difference between problems we don't know how to solve and problems we know how to solve but whose solutions we don't like. Permanent, economically rational carbon sequestration is the first kind; carbon emission is the second. Another example of the second kind is automobile travel speed. Reducing the average speed of cars in motion would be environmentally valuable, because it would reduce per-mile fuel consumption while simultaneously reducing the productivity of driving. (Drivers would be constrained from directly turning increased fuel efficiency into increased consumption because getting to their destinations would take longer.) In the United States getting drivers to slow down is generally viewed as impossible—didn't Nixon try in 1973?—but it's actually quite easy: all you need is a monitoring system that automatically imposes fines. During my visit to Australia in 2010, I marveled that traffic around me was moving at exactly the speed limit—then noticed the cameras mounted above the roadway. In the United States, such a system would be viewed as unsporting, among other things, and would lead to a ferocious public outcry. But it would work, and we know how to build it right now.

- The Kyoto Protocol was first adopted in 1997. A decade later, rounded to the nearest whole percentage point, the proportion of global energy consumption supplied by wind and solar, combined, was zero.

Their share has grown rapidly since then, as has the share of other renewables, but the absolute numbers and percentages are still very small. And the world's total energy consumption has grown, too.

Imagine that the world, tomorrow, reaches a consensus on energy and climate and rapidly does whatever needs to be done to slash global consumption of fossil fuels and hold atmospheric greenhouse gases at their current level. Imagine, further, that those measures are immediately effective and that average global temperatures stabilize, glaciers stop melting, and sea level remains where it is. Such a worldwide feat, however accomplished, would necessarily involve huge, permanent reductions in consumption—but imagine, as well, that those sacrifices are willingly made and are distributed equitably, so that the largest burdens fall not on the poorest but on those who, historically, have gained the most from human profligacy.

Now imagine what might come next. How likely would the 9 billion human residents of the world be, in the absence of any signs of worsening climate stress, to permanently endure, decade after decade, the continuing sacrifices required to maintain the new status quo—the halted growth, the forgone consumption, the reduced mobility, the population control, the willing abandonment of vast known reserves of fossil fuels?

At a talk I gave in 2011, I suggested that the Tea Party be renamed the Green Party because, by forcing spending cuts that seemed certain to prolong the recession and weaken the U.S. economy, it had done more to shrink America's carbon and energy footprints than any environmental group. I got the laugh I'd been hoping for, but somewhere in there was also a serious point. The yawning federal deficit could be thought of as a potential carbon sink, because paying it down would slow or halt U.S. economic growth and, by doing so, reduce the severity of a range of consumption-related environmental impacts. But attacking the deficit the way the Tea Party wants to do it, by cutting programs and policies that protect low- and middle-income citizens while maintaining those that serve the wealthy, would be a societal disaster. The only way to make true carbon-and-deficit reduction humane for the United States as a whole would be to finance it in ways the Tea Party wouldn't care for—say, by returning to 1950s-era income-tax rates, eliminating subsidies and tax breaks for the affluent, and—to keep the national focus from drifting—strengthening the E.P.A.

None of that is likely to happen, Warren Buffett notwithstanding—and the reason isn't only that right-wing ideologues would stand in the way. David W. Orr, who is the Paul Sears Distinguished Professor of Environmental Studies and Politics at Oberlin College, emailed me a broadly apposite passage from George Orwell's famous essay on Rudyard Kipling: "All left-wing parties in the highly industrialised countries are at bottom a sham, because they make it their business to fight against something which they do not really wish to destroy. . . . We all live by robbing Asiatic coolies, and those of us who are 'enlightened' all maintain that those coolies ought to be set free; but our standard of living, and hence our 'enlightenment', demands that the robbery shall continue. A humanitarian is always a hypocrite. . . ." Those of us who are "enlightened" on environmental issues earnestly wish for the world to consume and emit less—a radical transformation that we expect our own standard of living to survive.

Daniel Nocera, who is the Henry Dreyfus Professor of Energy at M.I.T, told me, "The fact is that we're crappy environmentalists. I started worrying about energy thirty years ago, but I haven't changed my lifestyle at all. Environmentalism is in our hearts, but not in our actions. Where we get confused is when we think that what's in our hearts takes care of the problem, when in fact it only takes care of what's on our consciences."

It's easy for wealthy people to look busy on energy, climate, and the environment: all we have to do is drive a hybrid, eat local food (while granting ourselves exemptions for anything we like to eat that doesn't grow where we live), remember to unplug our cell-phone chargers, and divide our trash into two piles. What's proven impossible, at least so far, is to commit to taking steps that would actually make a large, permanent difference on a global scale. Do we honestly care? That's the conundrum.